[**CHASTE MIMOSA**]

THE PSYCHOLOGY
OF PLANTS

ABOUT THE AUTHOR

Terence McMullen was born in 1937 in New South Wales, Australia. He obtained the degrees of Bachelor of Arts, with Honours Class I and Doctor of Philosophy at the University of Sydney. He lectured and supervised at all levels in the School of Psychology there, mainly in the areas of Personality and the History and Philosophy of Psychology. *Chaste Mimosa* is the product of hobby research, much of which was done in London, Dublin and Glasgow during Sabbatical leaves as an Honorary in Universities in those cities.

FIRESTICK ECOLOGY

FAIRDINKUM SCIENCE
IN PLAIN ENGLISH

FIRESTICK ECOLOGY

FAIRDINKUM SCIENCE
IN PLAIN ENGLISH

VIC JURSKIS

Connor Court Publishing

Connor Court Publishing Pty Ltd

Copyright © Vic Jurskis 2015, revised edition 2016.

PO Box 224W
Ballarat VIC 3350
sales@connorcourt.com
www.connorcourt.com

ISBN: 978-1-925138-74-0 (pbk)

Cover design by Maria Giordano

Front Cover: This ancient cypress stood sentinel to Major Mitchell's exploration of the Murray and his encounter with a war-party of Aborigines come over from the Darling on 27th May 1836. (Author's photo)

Back Cover: Ancient woodland of cypress on a sandhill above river red gum woodland, Lake Urana. (Courtesy NSW Department of Primary Industries)

Printed in Australia

CONTENTS

FOREWORD

I first came across 'plant ecology' as a young undergraduate student some 60 years ago. It was then rather an unglamorous subject, lacking the excitement of the rapidly evolving and breakthrough sciences like genetics, biochemistry and plant physiology. It seemed entirely qualitative; teaching was dominated by the European schools of vegetation description (for example, the Zurich-Montpellier school of vegetation description and classification) and by Clements in the United States and his stifling views on vegetation succession.

Clements' view was deterministic and 'holistic': "The developmental study of vegetation necessarily rests upon the assumption that the unit or climax formation is an organic entity. As an organism the formation arises, grows, matures, and dies ... each climax formation is able to reproduce itself, repeating with essential fidelity the stages of its development ... (The) climax is permanent because of its entire harmony with a stable habitat. It will persist just a long as the climate remains unchanged".

But Clements' insistence on a mono-climax led to a proliferation of terms describing the climax such as disturbance climax, edaphic climax, topographic climax, fire climax, even superclimax, and 33 other variations. Why all these terms? Because in trying to fit empirical evidence to a concept that insisted that all communities were but stages in progression to a mono-climax, the concept had to be bent, stretched or otherwise distorted to achieve the fit.

No wonder plant ecology seemed stolid and unscientific. Despite vigorous debate, the Clementsian view remained strong right up to the end of the 1900s. There really was no end to it until Paul Colinvaux wrote in his textbook in 1993: "No organising principles exist in nature to order the lives of plants in communities. All plants grow and reproduce in ways that promote individual fitness ... This is the mundane truth of the matter. And yet people tend to prefer brilliant myth to mundane fact ... The Clementsian myth of supernatural organisation in plant societies might require the hammering in of a sharpened stake before cock crow to end it".[I]

Why should Clements have held such sway and for so long? Clementsian ecology provided a semblance of pattern and order from which the field ecologist, presented with disorder, could find solace. Furthermore, it led to a whole suite of ecologisms such as pristine, self-sustaining, stable, harmonious, self-perpetuating – an ordered progress that is overwhelmingly appealing, but totally incorrect. New sorts of plant ecologies – quantitative ecology, ecosystem ecology, ecological physiology, theoretical ecology – began to emerge in the 1950s. For me, plant ecology became exciting. Ecologists began to measure how communities and ecosystems function, how they grow, their productivity, their responses, their interactions.

Now that we know more about how communities and ecosystems work, we recognise that plant communities are not in a stable state, living harmoniously with their environment. Rather, plant communities change in space and time according to climate and events that we have called 'disturbances'. An understanding of the role of these natural events in moulding community structure and function is therefore fundamental for the management of natural resources. For example, Jurskis shows that the state of many plant communities is now very different from that maintained by Aboriginal burning over millennia. I treasure the statement by American ecologist Orie Loucks: the elimination of disturbance by modern humans "will be the greatest upset of the ecosystem of all time … It is an upset which is moving us unalterably toward decreased diversity and decreased productivity at a time when we can least afford it, and least expect it".[II]

This book by Vic Jurskis addresses these last issues, particularly as they affect management, in a most readable and commonsense way. At the very basis of all of this perhaps are our notions of disturbance. Aber and Melillo wrote: "Disturbance by fire, defoliation, or other agents

I P. Colinvaux, *Ecology 2*. Wiley, New York, 1993.
II O.L. Loucks, "Evolution of diversity, efficiency, and community stability". *American Zoologist*, 1970, 10: 17-25.

is an intrinsic and necessary part of the function of most terrestrial ecosystems – a mechanism for reversing declining rates of nutrient cycling or relieving stand stagnation".[III] This is patently true – but then if disturbance is an intrinsic and necessary part of function, how can it be a disturbance?

While the emphasis of Jurskis' book is Australian, the relevance is global. For example, a recent paper from the USDA Forest Service shows that despite the huge arsenal and the billions of dollars spent annually in fire suppression, bushfires are becoming larger and more expensive, with greater damage to watersheds and communities. The authors conclude that the "recognition that fire exclusion is neither desirable nor possible in many regions of the country is an essential and … critical component of any effort to transform wildfire management".[IV] That sounds like commonsense to me and a little different from the view (slogan?) coming from one of the authors of a recent book published by CSIRO: "Fire ecology and management is not rocket science; it is much harder than that".

I hope that this book by Vic Jurskis will be widely read, and that it will play a valuable part in resolving on-going battles over the management and conservation of Australia's native ecosystems.

Peter Attiwill
Melbourne
18 May, 2015

III J.D. Aber and J.M. Melillo, *Terrestrial Ecosystems*. Saunders College Publishing, Holt, Rinehart and Winston Inc, Orlando, Florida, 1991.
IV D.E. Calkin, M.P. Thompson, M.A. Finney. "Negative consequences of positive feedbacks in US wildfire management". *Forest Ecosystems* 2: 9. 2015. *Forest Ecosystems* is an open-access journal, and this paper can be downloaded at http://www.forestecosyst.com/content/2/1/9

DEDICATION

Why does everything you know, and everything you've learned, confirm you in what you believed before? Whereas in my case, what I grew up with, and what I thought I believed, is chipped away a little and little, a fragment then a piece and then a piece more.

Cromwell's thoughts of More. *Wolf Hall*, Hilary Mantel.

This book is dedicated to the memory of Roy Free. During his distinguished career, Roy was Chief of Hume-Snowy Bushfire Prevention Scheme, Bush Fire Councilor, Chief Inspector of Forests in New South Wales, Assistant Commissioner of Forests, and a bloody good bloke. He was awarded the N.W. Jolly Medal by the Institute of Foresters of Australia for outstanding service to the profession.

Roy hired me as a labourer when I joined the Forestry Commission after graduating from university in 1976 because I needed to learn that academic qualifications count for nothing without hands-on experience. He explained this to me after he gave me a start as a professional forester in 1978.

Thirty years later, at an Australia and New Zealand Institute of Foresters conference in Coffs Harbour, Roy complimented me for "telling it how it is". This was after I had criticised the Council of Australian Governments' whitewash over the disastrous 2003 alpine fires that burnt 3 million hectares and ravaged our National Capital. In question time, COAG collaborator Professor Peter Kanowski claimed that "the jury's still out" on the efficacy of prescribed burning, but fire scientist Phil Cheney 'jumped in' to set the record straight. These pragmatists encouraged my campaign for improved land management in Australia, and I'm still trying.

Roy was always friendly and down to earth, notwithstanding his 'high office'. He was instrumental in developing firestick forestry. Until we are blessed with as great a man to champion firestick ecology, we will continue to suffer environmental and socio-economic calamity as a consequence of uncontrollable wildfire.

INTRODUCTION

There is a remarkable peculiarity in the trees in this country; however numerous they rarely prevent you tracing through them the whole distant country.

John Glover

Australian Aborigines hold the world record for sustainable development. They occupied Australia about fifty thousand years ago and developed the most durable culture the world has ever seen. Initially they created havoc, changing the pattern of vegetation across the landscape and exterminating the megafauna. After a few millennia they had established a new balance of nature that lasted until the late-eighteenth century invasion by Europeans. Aboriginal Dreamings refer to 'prehistoric' events evident to modern science such as human colonisation of Australia with the arrival in Arnhem Land of an ancestor of the Gagadju (with dilly bags of yams and a womb full of babies) or of an ancestor of the Yolngu from far across the sea in a canoe (with his two sisters); the post-glacial separation of Kangaroo Island from the mainland; and the creation of some crater lakes on the Atherton Tableland by volcanic eruptions ten or twelve thousand years ago.[V] The culture survived extreme climate change, hugely rising sea levels and volcanic traumas but it could not resist the European invasion.

In the Western Deserts, where European culture and technology were not capable of supporting human habitation, Aborigines continued to operate sustainable economies until they were mostly rounded up to make way for the British nuclear weapons testing program in the 1950s and 1960s. Probably the last exponent of fairdinkum sustainable management was Warlimpirrnga Tjapaltjarri who walked out of the desert in 1984.[VI] Instead of learning from this most durable culture, many Australian ecologists deny its power and work to extinguish the firestick forever. They start from a wilderness philosophy, variously described

---

V Josephine Flood *The original Australians*. Allen & Unwin. 2006.
VI Paul Toohey "The Last Nomads". In *The Bulletin*. 4 May 2004. pp. 28-35.

as naturism, bioaltruism, biocentrism or ecocentrism, and try and force nature to fit the philosophy.

If we don't recognise man as a part of nature we can't manage sustainably. Landmark works by Stephen Pyne and Bill Gammage set straight the historical record of Aboriginal fire management as it was in 1788. But Pyne was ambivalent, and Gammage was mostly silent about how the first Australians originally shaped the environment 50,000 years ago. Gammage portrayed a complex, spiritual view of Aboriginal burning. My story puts Australia's prehistory and history in a scientific context. It is a simple story of how Aborigines forged an economy out of the wilderness, how Europeans disrupted the ancient balance, and how we can restore it. I don't pretend to present a thorough scientific review. Indeed I hope that this overview of my experience with Australian ecology will persuade people that much of the modern literature is unworthy of review because it is based on false assumptions rather than observations, thinking and testing. Such literature is propaganda, not science.

Fire management is essentially a physical science. The responses of plants and animals to fire are governed by their physical characteristics in relation to the physical characteristics of each fire. Ecologists lacking knowledge of physics, and experience of fire behavior, have hijacked Australian fire management. They claim that frequent burning will extinguish some plants that must restart from seed after they are killed by fire. But whether plants can reshoot after their green shoots are killed or whether they must start again from seed is not so important. What matters is whether a fire is severe enough to kill mature plants, or only so mild as to kill small seedlings.

For more than 40,000 years, Aborigines lit mostly mild fires that consumed dead wood and dry herbage, killed seedlings, reduced some saplings back to ground level and scorched the leaves of some small trees and large shrubs. These fires allowed annual herbs to germinate and perennials to flush with new growth. They recycled nutrients from dead and dry into new growth. They maintained a diverse, open, safe and productive environment. After European settlers disrupted Aboriginal burning, dead wood, litter and dry herbage accumulated. Too many seedlings and saplings grew into trees or bushes. Delicate herbage was smothered and nutrient cycling was disrupted. Megafires, chronic decline of eucalypts, scrub invasion, pestilence, loss of biodiversity and socioeconomic distress have been the result.

A healthy and vibrant environment and economy can be restored if we dismantle the prevailing green academies and bureaucracies and reinstate fairdinkum science as the basis for managing our resources. Firestick ecology can win us back the *Magic Pudding*.

1

SCIENCE

Yes, how many times can a man turn his head
pretending he just doesn't see?
Bob Dylan – *Blowin' in the wind*

I attended a religious, boys-only secondary school where my earliest
science teacher was a remarkable woman. Doc Buhler (I didn't know
her given name) had been involved in the Belgian resistance against
Hitler's Nazis during the Second World War, and had gained a PhD from
the Sorbonne. She was a feminist who valued dual Nobel laureate Marie
Curie as a role model, but she also celebrated the different qualities of
the sexes. (Coincidentally my 'great uncle' Alfonsas Jurskis, who initiated
radio broadcasting in Lithuania, studied briefly in the department headed
by Maria Sklodowska-Curie in Paris.) Doc Buhler occasionally quoted
Einstein on God in a way that suggested her view was not exactly in
tune with the school administration. Recently I read a book by Richard
Dawkins[1] that takes a similar tack. It reminded me of my youthful naivety,
when I asked what I thought were innocent questions during discussions
at a religious 'retreat' and provoked the ire of the pastoral authorities.
But this reaction helped me to develop an appreciation of science, and I
am grateful for the education.

Doc allowed myself and some mates to develop our rudimentary
driving skills in her old Volkswagen 'beetle', and introduced us to the
pleasure of drinking rosé, but the thing I remember most was her attitude
to higher education. She told us that the role of universities was not to
impart knowledge, rather it was to train people to use their brains – to
think for themselves!

When I attended university I found that many of my lecturers
didn't share this attitude. They took a more pastoral view similar to the

religious at school. Nevertheless their attitudes reinforced my respect for independent thinking, and fortuitously immunised me against the current climate hysteria. After graduating as a forester I became involved in pseudoscientific political debates about management of public forests, quickly finding that Australian academia was turning out a lot of sheep conditioned to mill around with the mob. I came to realise that post graduate studies very often involve experiments designed to support conventional wisdom by using elaborate statistical procedures to force data into the mould. Alternatively, where there are no preconceived results, statistical methods are often used to search for answers rather than to test hypotheses that have been constructed on observations.

A memorable event occurred when a new syllabus was introduced at high school and I began to study physics. The course relied on a text whose name I have long forgotten but whose opening message was indelibly fixed in my memory. It introduced the scientific method using a parable of a Martian landing on earth with no source of heat for cooking. He observed a log fire, and hypothesised that cylindrical objects burn. When he tested the hypothesis using (I think) cast iron and reinforced concrete drainage pipes, he was forced to modify it, and eventually found by trial and error that wooden objects burn. Adapting this parable to explain modern Australian ecology and climate science, I would have the Martian using nuclear fission to demonstrate that his original hypothesis was sound.

Observation is the first essential step in the scientific method and thinking is the second. Proposing an hypothesis is the third, and testing it the fourth step. Unfortunately, modern research often begins at the fourth step by testing a preconceived hypothesis or, just as bad, bypasses the scientific method and uses data collection and statistical gymnastics to search for insights into perceived problems. This invariably gets people into trouble because they focus on association and neglect logical cause.

In recent years I have been embroiled in debates about the role of fire in Australian ecosystems. It never ceases to amaze me that whereas pragmatic managers embrace the scientific method of observation, hypothesising and testing, many self-professed scientists disdain the use of observations, disparaging them as 'anecdotal evidence'. They don't bother to think, and they embark directly upon experimentation and statistical gymnastics to prop up pre-conceived theories, chiefly that burning will stuff-up the environment. But this issue deserves its own chapter (Chapter 8).

Scepticism is integral with hypothesis testing, so the ongoing acrimonious debate about climate change has little to do with science, and scientific consensus is an oxymoron. But too many books have already been written on this subject to justify much further effort on my part. Historical climate change was more extreme during the Little Ice Age and the Medieval Warm period than it has been at any time since the Industrial Revolution, and despite sustained increases in carbon dioxide emissions no corresponding trend in temperatures has been demonstrated. I suggest that even the mildest sceptic would rightly discount the *Angry Summer of 2013*, claimed by the Australian Bureau of Meteorology to be the hottest Australian summer on record, and contrived by discarding all temperature data collected prior to 1910.[2]

Industrial emissions may well be having some effect on global temperatures, but the limited scientific component of the climate debate suggests that it is either very small or not distinguishable amongst all the more important factors, with their great variability. Modelling cannot resolve this debate because it cannot satisfactorily incorporate all the medium and long term cycles and interactions that are known to have a major influence on climate. These comprise several Milankovitch cycles in response to variation in Earth's orbit, rotational axis and wobble; other variations in accordance with solar activities including sunspots, flares and magnetic fields; and oscillations according to interactions between climate cycles and ocean currents, including the El Nino Southern Ocean phenomenon. Furthermore, there have been very abrupt climatic changes such as the Younger Dryas that are not fully understood.[3] Scientific observations of the earth's climatic prehistory and history do not support an hypothesis that man is causing dangerous global warming, and such an hypothesis isn't currently testable.

But some hypotheses can be tested against further observations without resorting to conventional experimentation, and this can be a very efficient process because a single contrary observation can forestall time-consuming and expensive experimentation and statistical analysis. New South Wales Forestry Commissioner in charge of research (amongst other things), Jack Stewart, was renowned for summing up field inspections of the outcomes of contrasting approaches to forest management in the following terms: "you don't need a fucking chi-square test to see the difference". When statistician Bob Gittins later explained to me that "statistics do not provide answers to questions, they merely give confidence in the limits of data", I remembered that Jack had put it more clearly and succinctly. Sound hypotheses developed from careful observations without prejudice can satisfy open minded thinkers without any fancy statistical support. Darwin's (or maybe Wallace's[4]) theory of evolution of species by natural selection, or survival and reproduction of the fittest, was constructed without experimentation or number-crunching, has mostly (see Chapter 13) stood testing against further observations over time, and is supported by high-tech new specialties such as molecular genetics.

Experimentation and statistics are not, as many academics would have it, an essential part of the scientific method. Rather they often provide distractions. This became apparent to me through long and unproductive debates with forest pathologists about chronic eucalypt decline (Chapter 6). I was condescendingly lectured on many occasions about the need for "empirical rigor" by pathologists and entomologists who had not observed the process as I had, and did not understand it. They wrongly equated empiricism with experimentation. According to the Oxford Dictionary, empirical means "verifiable by observation or experience rather than theory". However these pathologists disdained my observations and experience. They adopted illogical theories, used statistical gymnastics to demonstrate association, and thought that they had proved cause.

Any thinking person should easily understand the difference between proof of association and proof of cause, but many academics and bureaucrats do not. For example, warm and wet conditions coincided with the release of myxoma to control rabbits in 1950. These conditions favoured its spread by mosquitoes and also favoured the spread of human encephalitis by mosquitoes. Some public health officials thought that the newly released myxoma virus was likely responsible for the coincidental outbreak of encephalitis. Professor Frank Fenner and Sir McFarlane Burnett injected each other with myxoma to allay public fears.[5] In a more recent example, scientists investigating rural tree decline in Tasmania's midlands attributed it to intensive grazing. The cause of eucalypt decline in this case was actually pasture improvement – the sowing and fertilising of leguminous, nitrogen fixing pastures to support intensive grazing (Chapter 9). Even worse, an association of bellbirds with chronic decline of forests has been tendered by forest pathologists and entomologists as evidence that colonies of bellbirds are the cause, when they are actually the result (Chapter 6).

According to my high-school physics text, parsimony is an important component of science. Unfortunately I can't recall how that was illustrated by parable. Along with observation and independent thinking, parsimony is seriously undervalued in much Australian ecology, which seems to glory in complexity. A prime example is the concept of "eucalypt diseases of complex etiology"[6] whereby every region, forest type and pathological specialty seems to need its own particular and complex explanation for the widespread and simple issue of chronic eucalypt decline (Chapter 6). My colleague Frank Batini in Western Australia doesn't like what he calls my "one size fits all" approach to this issue, but he's a specialist. Frequent and deliberate use of Occam's Razor could save Australian taxpayers billions of dollars that are currently spent on very specialised, sometimes poorly designed and, too often, inconsequential research of relatively simple problems.

For example, in 2012, the Centre of Excellence for Climate Change, Woodland and Forest Health in Western Australia comprised two universities, a State Government department, twenty seven collaborating industry partners, seven collaborating academic institutions and four programs run by seven program leaders. The Centre had focused on decline of tuart (*Eucalyptus gomphocephala*) and wandoo (*E. wandoo*) woodlands for three years without actually proposing any solutions, and was rewarded with funding for an additional three years to study decline of marri (*E. calophylla*). These kinds of arrangements are a great incentive for researchers to strive for complexity and specialisation rather than practical solutions.

A good example of the use of scientific method compared to a more typical modern approach is available from two studies of the habitat preferences of smoky mouse, a rare and endangered species, in southeastern New South Wales. The first study commenced after a live specimen had been trapped for the first time in this State, in ridgetop heathland near Eden during the spring of 1994. We (myself and colleagues at Eden Forest Research Centre) observed that five quite different types of potential habitat were available around our capture site and hypothesised that the mouse would preferentially forage in one or another of the potential habitats. So we used a stratified random sample and a low density of traps across 30 hectares around the site to test the hypothesis.

We found that:

> *Pseudomys fumeus* was trapped in rocky, exposed positions with low densities of ground and shrub cover. It was also trapped in an adjacent area which had been logged and burnt two years previously. This area was not rocky and exposed. … One species of plant, *Kennedia rubicunda*, occurred in all vegetation plots associated with *P. fumeus* captures and no vegetation plots where *P. fumeus* was not captured.[7]

We next hypothesised that there may be a particular food associated with bare and rocky or disturbed ground that is important to the mouse during spring, and nominated *Kennedia* as a potential example because it develops large legumes with many seeds at this time of year, and ripening legumes are a rich source of protein. We suggested that this hypothesis and any seasonalilty of habitat use could be tested using a combination of dietary and behavioural studies.

Alas, on the usual assumption that human activity such as logging or burning is a threat to wildlife, our study area and the rare mouse were 'protected' in a new national park. The parks authority engaged a postgraduate student to study the habitat preferences of the mouse. Disdaining our observations and hypotheses, this student duplicated the methods from a twenty year old study on a heathy ridgetop 700 kilometres to the west, in the Grampian Mountains. One of his supervisors was the author of the previous study and unsurprisingly the conclusions were very similar. "Multiple logistic regression of habitat variables and capture locations revealed a floristically determined preference for heath habitat".[8] *Kennedia* was rare in the allegedly preferred floristic type.

The student had placed a high density of traps in a small area (7.5 hectares) around the heathy nesting sites of the smoky mice. They could not help but be attracted to the aromatic baits before they had any opportunity to display their natural foraging preferences. The unfortunate consequence was that lactating females apparently suffered a nutritional crisis and the local population became extinct by midsummer. (Smoky mice are very short-lived, with males surviving a single year, and females one or two.) The resulting scientific article suggested that:

> Causes of the decline were unclear. … The sudden decline in the study population, combined with the patchy distribution of suitable habitat and a high level of male transience, suggests that *P. fumeus* form a metapopulation in the Nullica region.

I have little doubt that the population was extinguished because breeding females were attracted to the baited traps nearest to their nests and were thus denied a critical resource, most probably *Kennedia* seeds, which grew further away. In my view, the only meaningful data produced by this student's work were descriptions of the plant communities growing on sites where the mice were able to excavate burrows and hide from diurnal predators.[9]

Whether the mice possibly foraged in different habitats well outside the ridgetop heath wasn't tested. The study assumed that the mice foraged in heath, and unintentionally modified their natural behaviour by providing an abundance of attractive, exotic food that 'confirmed' this assumption. Unfortunately the two widely separated populations of this rare and endangered mouse may have been similarly affected by interference with their foraging behaviour. The article reported that "captures of *P. fumeus* in Victoria have become rare, even from sites such as Mt. William (A. Cockburn unpublished data), long held to be the stronghold of the species".

Like much modern ecology, this study started from assumption rather than observation, resulting in an inappropriate design, and used statistical gymnastics to look for an answer rather than testing an hypothesis. Thus smoky mice became casualties of 'science'.

Endnotes

1 Dawkins 2006.
2 Australian Government Bureau of Meteorology 2012, 2013.
3 Carter 2010.
4 Wallace 1858.
5 Australian Government 2002.
6 Old 2000.
7 Jurskis *et al.* 1997.
8 Ford *et al.* 2003.
9 Jurskis 2004a.

2

ECOLOGICAL HISTORY

The essence of man as a rational being is that he develops his potential capacities by accumulating the experience of past generations.

E.H. Carr

History and prehistory are the richest sources of observational data available to science. Unfortunately, many Australian ecologists uncritically dismiss this wealth of evidence because it does not conform to their preconceptions, particularly about Aboriginal burning and vegetation change (Chapter 4). For example Dr. Philip Gibbons and his colleagues at Australian National University's Fenner School of Environment generalised that explorers' accounts were subjective, qualitative, unsystematic and influenced by pecuniary interest. They made the incredible assumption that areas without any treecutting, firewood collection, grazing by domestic stock or recent fire were representative of pre-European conditions[1] despite the historical evidence that Aborigines felled trees and burnt wood frequently and widely in broadcast and spot fires that maintained the open grassy ecosystems sought by European pastoralists.

Noted fire ecologist Professor David Bowman claimed that "It is remarkable how little is known about the use of fire by Aboriginal people. ... Perhaps the greatest limitation of historical records is our inability to determine whether Aborigines had a predictive knowledge of the ecological outcomes of burning".[2] In other words, Bowman suggests that we dismiss the comprehensive historical records because we're not sure if Aborigines knew what they were doing after forty thousand years of collective experience in 'firestick farming'. Dr. Matthew Colloff argued that "the subjectivity of historical accounts is a real issue for ecologists".[3] He cited Bowman's conclusion that "Given the rapid loss of cultural knowledge from so many Australian landscapes we are forced to puzzle over the past".

Colloff puzzled unnecessarily over the history of river red gum (*Eucalyptus camaldulensis*) forests (Chapter 11). He wrote that: "Reservations by ecologists about historical accounts of Aboriginal fire focus on issues of cause and effect, separating fire as one factor among several. … What about where flooding was the primary force structuring the landscape?". Actually, it's not at all difficult to separate these factors because Aboriginal burning on the Murray was disrupted from 1842, and the river red gum forests subsequently grew up under natural flooding regimes. Separate impacts of river regulation have been apparent since the 1920s and 1930s when major infrastructure dramatically changed flooding regimes. There is no mystery about the historical interaction of flooding and Aboriginal burning, or the impacts of European settlement (Chapter 11).

Denial of history is well entrenched in an 'Old Boys' Club' of Australian ecologists where they don't appreciate the grit in the oyster, such as that inserted by Pyne[4] and Gammage,[5] and are therefore denied any pearls of independent wisdom. In 2012, with my colleague Roger Underwood, I submitted a comment based on ecological history to *Austral Ecology*. The editor considered that we made a strong and well argued case but rejected it on the basis that they no longer published comments on individual papers. We widened the scope to a mini-review, but this time the editor rejected it because our argument was allegedly not strong enough to "swim against the tide". One hostile reviewer referred to direct observations by Colonial Surveyor-General Major T.L. Mitchell that we cited in our manuscript, as "erroneous inferences". Our paper has since been published in *Fire Ecology*,[6] an international journal which encourages scientific debate.

A wealth of reliable information on Aboriginal Australia was collected by officers and servants of the various colonial governments and by other explorers, naturalists and anthropologists with no 'pecuniary interest'. This was usually collected systematically and objectively, and reported in their journals. Though it is mostly qualitative, it is real, unlike the numbers collected by modern scientists from environments that are vastly different from those originally explored. There are also a great many historical drawings and paintings that show what the country was like. For example, Fig. 1 is a drawing of a grassy woodland made by Major Thomas Mitchell using a camera lucida to ensure accuracy. In my historical ecological investigations of southeastern Australia I have particularly valued the accounts of Watkin Tench,[7] Matthew Flinders,[8] John Oxley,[9] Charles Sturt,[10] Thomas Mitchell,[11] Charles Darwin,[12] various colonial foresters,[13] Edward Curr[14] and Alfred Howitt.[15] Many of these men observed and described the dramatic changes that occurred after Aboriginal burning was disrupted (Chapter 4), changes ignored by theoretical ecologists who measure benchmarks for imaginary pre-European ecosystems. Curr deserves fame as the first scholar to describe firestick farming – a concept usually attributed to modern anthropologist Rhys Jones.[16]

None of these explorers and naturalists were specialists in botany. Lest I stand accused of bias in neglecting the accounts of botanists such as Cunningham, whose observations figure prominently in at least one

Fig. 1 "A grassy valley" in the Brigalow Belt of Central Queensland, T.L. Mitchell, July 1846. (Courtesy Archive Digital Books Australasia)

historical study by Croft and colleagues,[17] I should explain my reasons. Oxley made it clear that Cunningham deliberately biased his samples of flora by concentrating on unusual and restricted habitats so as to find rare plants. For example, he remarked at one point during their explorations that "the scrubs were the only vegetation productive to the researches of the botanists". Croft's study showed that the frequencies of shrubs recorded in various parts by Cunningham were much higher than the frequencies recorded by five separate explorers in the same region. The shrubs mostly grew in restricted areas where they had a competitive advantage over trees, herbs and grasses (Chapter 3). These were the areas where Cunningham concentrated his efforts.

The explorers, on the other hand, traversed and described whatever lay between them and their ultimate goals. They routinely described what they encountered in ascending prominences to triangulate their maps. Thus they surveyed the restricted habitats that supported scrubs as well as the extensive open and grassy habitats that they sought for pastoral development. It is clear that the explorers accounts were objective and systematic compared to modern 'benchmarks' derived from small areas showing little sign of human activity.

Some time ago I challenged a paper comparing scrubby, unburnt and ungrazed forest with grassy, grazed and burnt forest, and purporting to show that grazing and burning depleted shrub understoreys.[18] The authors of this article were allowed to publish a response alleging that I misrepresented their data:

> In support of the case for shrub invasion, Jurskis appears to misrepresent our data. He states 'there was no correlation between adult and juvenile shrub densities'. For reasons outlined previously herein, we did not calculate such a correlation. It is difficult to see how Jurskis could have calculated a correlation without the raw data.[19]

However these authors had stated in their original paper "There was no significant correlation ($r = 0.31$ NS) between the densities of adults and juveniles among the woody species".[20] I brought this to the attention of the editor, but he declined to publish a correction and suggested I sort it out personally with the senior author of the response, who happened to be an associate editor of the journal. Ecological historians are indeed swimming against the tide of theoretical ecology.

Dr. David Keith has published other material showing a lack of appreciation of our ecological history, including a book about the native vegetation of New South Wales[21] that was commissioned by the National Parks and Wildlife Service. In this book he described many chronically declining, shrub-infested eucalypt communities as though they were natural ecosystems rather than artefacts of post-European mismanagement. These are now mostly listed as endangered ecological communities and grazing and fire are listed as threats. The consequences are that it is illegal to manage any remaining healthy stands properly or to restore more natural structure and function in the degraded scrubs. For example, Keith classified as "wet sclerophyll forest" (see Chapter 12), naturally grassy spotted gum (*Corymbia maculata*) forest that is turning to scrub because of reduced burning. Patches of this type of scrub at Pittwater near Sydney are now on the 'endangered list'.

Keith and some of his colleagues claimed that:

> there may be conflict between conservation goals and the 'traditional' use of fire on Aboriginal lands … Aboriginal fire regimes were instituted largely to maintain access to food resources, whereas conservation management may involve very different goals, such as protection of a rare plant species which may have been of no value to Aboriginal people and indeed may have been disadvantaged by Aboriginal burning regimes.[22]

But species that were disadvantaged by Aboriginal burning regimes are now either extinct like the megafauna, or they are currently escaping their long confinement to sites that are physically inaccessible to mild fire.[23] The latter species are choking out grassy woodlands and forests. These include the shrubs and vines that characterise Keith's "Southern Lowland Wet Sclerophyll Forest" – an unnatural scrub created by modern neglect.

Ironically, a prehistorian named David Horton explicitly pushed the idea that theoretical ecology has some scientific merit not possessed by history. Horton argued from ignorance of the relationships amongst intensity, frequency and extent of fire, and ignorance of the consequent impacts of different fire regimes on plants, animals and charcoal deposition (Chapter 8). He composed a litany of fauna that supposedly could not persist under regimes of frequent fire and claimed that their persistence, together with sustained low levels of charcoal deposition, indicated that Aborigines didn't do much burning.[24] Horton neglected to distinguish infrequent, high-intensity fires that kill all standing vegetation, and produce bulk charcoal, from frequent low-intensity fires that maintain open grassy vegetation with mature trees and shrubs, and produce mostly ash. Using a similar approach, Dr A. Malcolm Gill initiated a list of plants that is now used by New South Wales' bureaucracies to inhibit prescribed burning (Chapter 8).

Horton showed his lack of appreciation of history in suggesting that Tindale conceived the idea, in the 1950s, that fire was critical to Aboriginal economy. This concept was evident in the journals and publications of all the historical ecologists I have mentioned, and was clearly and explicitly espoused by Curr in 1883. Incredibly, Horton claimed that post-European changes in native vegetation were unrelated to disruption of Aboriginal management. Bowman showed a similar neglect of history when he stated that "It was not until the second half of the twentieth century that the fundamental role of fire in the dynamics of many Australian terrestrial ecosystems was recognised".[25] He cited modern ecologists Bradstock, Williams and Gill to support this revision of our past.

Reading this prompted me to search the net, and I found a blurb by CSIRO Publishing for a book titled *Flammable Australia – Fire Regimes, Biodiversity and Ecosystems in a Changing World*:

> It is edited by Professor Ross Bradstock from The University of Wollongong's Centre for Environmental Risk Management of Bushfires, Dr Malcolm Gill from ANU's Fenner School of Environment and Society, and Dr Dick Williams from CSIRO's Ecosystem Sciences. … 'This book is the third synthesis of its type

I have been involved with in the last 30 years', Dr Gill said. 'It asks
questions about enduring concepts and themes in fire ecology and
management, but also looks to the future, focusing on the effects
of global change on fire regimes. Fire ecology and management is
not rocket science; it is much harder than that'.

Ecologists such as these have certainly made a hard job of
straightforward science by ignoring history. As long ago as 1802, Matthew
Flinders recognised the differences between dense woody vegetation
in uninhabited wilderness compared with the mostly open, grassy and
frequently burnt character of mainland Australia. He deduced that
infrequent high-intensity lightning fires visited the wilderness whereas
human fires maintained the open and safe grasslands, woodlands and
forests typical of Australia when Europeans arrived (Chapter 8). Two
centuries later many fire ecologists remain unaware of this fundamental
distinction. Bradstock, Williams and Gill are still wondering "what causes
major fires" and "to what extent can we … influence fire regimes".[26]

A group of fourteen such ecologists, over a period of four years,
recently published three articles in scientific journals, under various
reshuffled combinations of between eleven and thirteen authors,
portraying extensive, dense, even-aged stands of mallee created by post-
European megafires (Chapter 8) as though they are a natural phenomenon
that existed under Aboriginal management.[27]

Thirteen of them stated that:

> Mallee vegetation has a low canopy (generally <5 m), and so almost
> all fires (prescribed or wildfire) result in the canopy burning and
> the death of above-ground stems and branches. New stems then
> arise post-fire from synchronous resprouting from the lignotubers.
> Consequently, most mallee stems in a particular location are the
> consequence of a single fire event.[28]

This misleading statement was unsupported by any reference to other
scientific literature, and was contradicted by an admission, in a later
paper in the series, that multiaged stands were deliberately excluded from
sampling.[29] Nevertheless it survived peer review and was published in

Australian Journal of Botany. It became the foundation for papers arguing against prescribed burning of mallee. They claimed that long-unburnt stands are the only reservoir of faunal habitat features such as holes in trees that take a long time to develop.

Had these scientists acquainted themselves with our ecological history, they may have argued instead for more intelligent prescribed burning of mallee so that young trees could grow old. Surveyor-General Mitchell noted of *Eucalyptus dumosa* that "their disproportioned roots also prevent the bushes from growing very close together ; and the stems being leafless, except at the top, this kind of eucalyptus is almost proof against the running fires of the bush".[30] Pioneering pastoralist Edward Curr wrote that:

> One of the articles of food which we found most in use amongst the Blacks of the Lake Boga neighbourhood and the mallee scrubs was manna.[I] There were bags full of it in almost every camp, and I understood the Blacks to say that they used to set fire to a portion of the mallee each year and gather the manna the next season from the young growth.[31]

Aborigines mostly burnt the mallee with low-intensity fires that sometimes scorched the canopy but usually did not consume it as do modern megafires. Dormant bud strands in branches sprouted and produced abundant soft and nutritious foliage which fed large numbers of bugs. These produced a bounty of manna for Aboriginal people. Aboriginal fires were confined to small portions of the mallee. Now "almost all fires (prescribed or wildfire) result in the canopy burning"[32] because dry leaves, bark and twigs of mallee as well as dense rank stands of porcupine grasses (*Triodia*) are allowed to accumulate for decades before they are burnt either deliberately or accidentally. Over 35 years spanning the turn of the millennium, infrequent (> 35 years between fires) megafires (> 10, 000 hectares per fire) ravaged the Murray Mallee Region[33] because heavy three-dimensionally continuous fuels inevitably produce uncontrollable crown fires, whatever the height of the canopy (Chapter 8).

I Sugary exudates produced by bugs feeding on the sap of trees.

During the 1980s, CSIRO trials had demonstrated that frequent burning could restore grasslands in swales which had been invaded by mallee and other woody plants.[34] These grassy swales between mallee covered dunes were invaded by trees and bushes after Aboriginal burning was disrupted. Rather than arguing against burning, modern ecologists should be arguing for this kind of more intelligent burning to restore biodiversity, fuel diversity, fire safety and old trees on the dunes.

Ironically, in 2014, another ecologist was lamenting that extreme frosts, supposedly due to global warming, were eliminating mallees and banksias from swales. Professor Ian Lunt claimed that the banksias were attuned to fire but not to climate change.[35] These woody plants are actually attuned to living at low densities on dunes where fires are naturally less frequent than in swales because grass doesn't grow so profusely. Seedlings germinating after fires have a chance to grow to a size where they can withstand low-intensity fires. Frosts are never as severe on the dunes because cold air drains down into the swales on still, frosty nights. The woody plants didn't naturally grow in the swales where fires were frequent and frosts are naturally more severe. Natural climate extremes can help to restore more natural vegetation patterns (Chapter 5).

Past Australian of the year – Tim Flannery – used evocative quotations from Mitchell and others to describe Aboriginal fire regimes and the drastic changes in vegetation when they were disrupted (see Chapter 4). Unfortunately, Flannery didn't appreciate the drastic changes that had occurred when they were first applied. It is well established that Aborigines arrived, fire increased and megafauna became extinct during the same period of time, but there is no evidence of Aborigines hunting megafauna. According to the scientific method, these observations would not inspire an hypothesis that people hunted the megafauna to extinction. Flannery's motivations are obscure, but it's worth examining his arguments in view of the popularity of *The Future Eaters*,[36] as well as his high profile as Australian of the year in 2007 and former Chief Climate Commissioner for the defunct Gillard/Rudd – Brown/Milne Government.

Flannery suggested that the restless earth has destroyed the evidence of hunting. One wonders why there remains separate evidence of megafauna and Aborigines but not of their interaction. Flannery explained that if the megafauna succumbed quickly like the moa in New Zealand, it left a very narrow window of time for archaeological deposits. I don't buy that, because a large amount in a short time makes for a very obvious deposit as any money-laundering criminal would realise of bank deposits. Indeed Flannery reported that there are hundreds of Maori cooking sites containing the remains of moa across the length and breadth of New Zealand. Individual sites contain the remains of up to tens of thousands of moa. If Aborigines had eaten the Australian megafauna, there would surely be evidence.

To emphasise the potential impacts of hunting, Flannery gave a discourse on 'time dwarfing'. Fish supposedly got smaller because fishermen killed the big ones and let the little ones breed. Kangaroos, wallabies and koalas similarly got smaller because of Aboriginal hunting pressure. As proof of this pressure, Flannery referred to historical evidence that koalas and possums proliferated after European settlement released the pressure. But these animals weren't responding to reduced hunting as originally suggested by observers such as Rev Peter MacPherson in 1886 or Harry Parris in 1948. Flannery and other modern authors such as Roger Martin and Tim Low have merely regurgitated these speculations without critically examining the ecological history.[37]

Current 'overbrowsing' by koalas in Victoria and Kangaroo Island, and by possums in Tasmania's Midlands are repeat performances. When Aboriginal burning was originally disrupted, and more recently when grazing and burning were excluded from new national parks and hobby farms, soils became less healthy for eucalypts. Sowing pastures in woodlands has a similar effect on soils and eucalypts. Declining trees lose their chemical defences against everything that eats them, including koalas and possums. Their leaves become much more palatable and nutritious, so the marsupials flourish (Chapter 6).

Anyway, why didn't the megafauna just shrink? Flannery says Aborigines ate them all because they were large and slow moving targets

– easy to see and easy to hit. But Aborigines ate lots of possums. They caught them easily by climbing trees and smoking them out of hollows (Chapter 3). How did they survive? Flannery says koalas and wombats (and presumably possums) survived because they can hide in trees or burrows, but the megafauna couldn't hide so they were all caught and eaten. In fact, Aborigines burnt out soft-leaved trees and bushes,[38] so the megafauna had nothing to eat. Koalas and wombats survived because they were already adapted to poor quality forage, and plants that are adapted to, indeed dependent on fire, are poor quality forage. Poor soils, hard leaves and frequent fire all go together (Chapter 6).

Flannery claims that Aborigines couldn't have increased the impacts of fires because there were already plenty of lightning strikes, and increasing the frequency of fires lowers their intensity. Superficially this is plausible but only because Aborigines made it so by burning the megafaunal browse and creating open grassy woodlands where fine fuels such as cured grasses, dry leaves and shed bark can be burnt under mild conditions (Chapter 8).

Dense damp forests had greeted the first people to come to Australia. Lightning strikes rarely started running fires, but when they did, they developed into firestorms. Matthew Flinders described the vegetation on the uninhabited Kangaroo Island:

> A thick wood covered almost all that part of the island visible from the ship; but the trees in a vegetating state were not equal in size to the generality of those lying on the ground, nor to the dead trees standing upright. Those on the ground were so abundant, that in ascending the higher land, a considerable part of the walk was made upon them. They lay in all directions, and were nearly of the same size and in the same progress towards decay; from whence it would seem that they had not fallen from age, nor yet been thrown down in a gale of wind. Some general conflagration, and there were marks apparently of fire on many of them, is perhaps the sole cause which can be reasonably assigned.[39]

The first Aborigines deliberately burnt anything that was flammable.

They worked from sunny and grassy edges, gradually pushing fires further into forests and changing their character. After lightning strikes and firestorms felled damp forests, Aborigines burnt the dead wood, litter and regenerating seedlings as soon as there was enough dry fuel to carry fire. Thus they locked wet forests and fire-sensitive plants into small cells that were nearly always either too damp and sheltered, or too bare to carry fires (Chapter 12a). Flannery argued that fires increased because there was more wood to burn after newly arrived Aborigines ate the megafauna that previously ate the biomass. But the megafauna didn't eat wood. It seems to me that Flannery's arguments strive to put man's 'interference' with nature in the worst possible light. Perhaps exterminating the megafauna by eating them is more satisfyingly culpable in a naturist philosophy than exterminating them indirectly.

Australia's original human immigrants gradually burnt out the soft-leaved trees and bushes that sustained the megafauna, and in so doing produced much charcoal. Fire-tolerant trees and grasses prospered. After that, seasonal curing of tufts of grass, and of the litter from hard-leaved trees allowed Aborigines to light mild fires in summer. They were slow moving fires in the litter of the forest floor and fast moving fires in grassy woodlands. These fires produced much ash and little charcoal.

However, scientific debate about Aboriginal impacts on the environment took a curious turn in 2011 when an article based on prehistoric charcoal records was published in *Quaternary Science Reviews* by Dr. Scott Mooney with 18 colleagues.[40] This article reached conclusions that appeared to be at odds with the data presented, and included some perplexing analyses. The overall approach was to aggregate a large number of temporally and spatially dispersed samples of charcoal deposition around which there was individually much difference in interpretation. Mooney's paper argued that aggregating the data allowed them to 'resolve" some persistent controversies about Aboriginal burning.

The conclusion was that there was "no distinct change in fire regime corresponding with the arrival of humans in Australia at 50 +/- 10 ka" (thousands of years ago). However, the analysis showed two distinct peaks of charcoal deposition (i.e., evidence of increased burning of wood) immediately before and after 40 ka at two different time scales of variability (Mooney's Figure 2). The coincidence of human arrival and peaks in biomass burning is well within the bounds of reliability of the various dating methods. The existence of two peaks provides some small support for speculation by archaeologist Josephine Flood that Aboriginal ancestors were originally alerted when they saw the smoke from wildfires hundreds of miles across the sea in Australia.[41]

I queried the conclusion with one of the coauthors who has previously published a contrary view. It turns out that they supplied some of the data for the analysis but had little influence on its interpretation. Mooney's unusually long list of coauthors possibly suggests a broad database more than a broadly-based resolution. Later on, Mooney with a very much shorter list of collaborators contributed a chapter to *Flammable Australia*. They reported a "very large increase in biomass burning" after European settlement, and crticised Pyne's historical narrative because it "emphasised deliberate fire suppression as characteristic of the European colonisation of Australia".[42] Fire ecology is 'harder than rocket science' because ecologists have consistently failed to grasp the simple fact that is so obvious to historians: Fire suppression never worked and never will. It inevitably leads to megafires, and megafires produce huge amounts of charcoal.

Mooney's paper argued that widespread burning by Aborigines would be evident in high levels of charcoal deposition. But this argument contradicts the statement that "all of the regions show reduced fire during the past ca 50 years, and this is despite the widespread use of prescribed burning". Furthermore, a rough index of human occupation from archaeological sites in the arid zone was compared against charcoal samples from the humid zone, and claimed as proof that there was no correlation between Aboriginal occupation and burning.[43] Obviously there would be no direct relationship between human population in one area and human activity in a different area, even if the indices were sufficiently precise to allow sensible analysis. They were not. But in any case, analysis of charcoal from sediment cores cannot resolve disputes about low-intensity fires, nor can counting fire scars on trees, because low-intensity fires burn little green wood. Thus, little charcoal is produced and mature trees are rarely scarred by mild fires.[44]

The data compiled by Mooney clearly showed that unusually large amounts of wood were burnt around the time humans originally occupied Australia, and again after European settlers reinstated some vegetation similar to that in pre-human Australia. Disruption of Aboriginal burning set the scene for megafires that delivered heaps of charcoal (Chapter 8). The original interpretations by palaeologists who collected the data typically admit a peak in biomass burning coincident with human arrival in Australia. For example "There is a notable increase in fire activity centred on 40 ka before present which, in the absence of a major climate change around this time, is considered to most likely indicate early Aboriginal burning".[45] They also admit a reduction in charcoal after burning created open grassy vegetation with sclerophyll trees and shrubs. Data from uninhabited Kangaroo Island support Matthew Flinders' deduction that the fire regime comprised infrequent high-intensity lightning fires. Sediment cores confirm that dense woody vegetation proliferated and charcoal increased after the demise of the local Aborigines following the massive rise in sea level that separated the Island from Mainland Australia.[46]

Pre-history shows that Aborigines used fire to change the pattern of vegetation in Australia and together with history shows that they maintained the new pattern for about forty thousand years. The megafauna disappeared as Aborigines created the new pattern and there is no credible reason to suspect that they ate all the big animals. Destruction of browse by Aboriginal burning can easily explain both the demise of the megafauna as well as 'time-dwarfing' of the remaining fauna.

Most scientists attribute reducing body size over time to nutritional stress with changing climate. Flannery prefers the hunting explanation because dwarfing continued as climate ameliorated during the current interglacial. However we know that Aboriginal burning virtually eliminated the nutritious soft-leaved plants that sustained the megafauna. Improving climate didn't make the surviving hard-leaved plants any softer, it merely modified their local distribution as softer habitats marginally expanded (Chapter 12a). The surviving fauna continued to eat what they were used to eating, and selection for compactness continued.

Since Flannery wrote his book, further evidence has emerged to disprove his hypothesis. In 2005, Gifford Miller and colleagues published data from prehistoric eggshells and teeth showing that emus and wombats survived on a poorer diet after Aboriginal burning changed Australia's vegetation, whereas another large bird, the extinct *Genyornis* (jaw bird) had a more specialised diet and was unable to adapt.[47] In 2010, Gavin Prideaux and colleagues found evidence of thousands of years overlap in human and megafaunal habitation of southwestern Australia and no evidence that Aborigines hunted megafauna. However there was clear evidence of predatation by carnivorous megafauna such as *Sarcophilus* (cousin to the Tasmanian devil) and *Thylacoleo* (the marsupial lion). A giant snake, *Wonambi*, named for the 'mythological' rainbow serpent disappeared at the same time. Cultural taboos and sacred sites associated with the rainbow serpent are consistent with lack of evidence that Aborigines hunted it. Prideaux and colleagues also confirmed that a large amount of biomass was burnt when man arrived in southwestern Australia.[48]

In 2012 Flannery published an essay ignoring this further evidence and restating his hypothesis. He referred to new research published in *Science* purporting to show that charcoal deposition in Lynch's Crater on the Atherton Tableland increased **after** the megafauna were extinguished.[49] In fact Sporormiella (a fungus that grew in megafaunal dung) decreased in proportion as charcoal increased in sediments. The change in composition of vegetation apparently lagged behind by a century or so. Susan Rule and colleagues argued that "This rules out fire as a cause of megafaunal extinction, because that hypothesis depends on fire causing vegetation change, which would then cause megafaunal decline. In that case, vegetation change should have led megafaunal decline, rather than following it".[50]

I suggest that "the extended trajectory of the rise in charcoal and its close matching with falling Sporormiella"[51] is proof positive that Aborigines quickly reduced the forage available to megafauna by burning the soft young growth of shrubs and small trees at the edges of the forests. It would take some time for repeated edge fires to gradually open up the canopy and penetrate into the heart of the forest, so that firstly grasses and then hard-leaved trees could gain dominance over soft-leaved trees (Chapter 12a). The megafauna would have starved long before the big old soft-leaved trees were finished off and grassy sclerophyll woodlands were fully established.

Megafaunal dung started to decline at the same time as Aboriginal burning started to produce more charcoal 42,000 years ago. About three hundred years later, maximum rates of increase in charcoal and decrease in dung were attained. The maximum rate of increase in charcoal was sustained for about two centuries, whereas the maximum rate of decrease in dung was only sustained for about a century.[52] In other words, accessible, soft browse for megafauna ran out before the woody material that Aborigines turned into charcoal.

Furthermore, fungal spores by their nature and purpose are quickly and widely dispersed, whereas there is a time-lag in delivery of charcoal to sediments by overland flow within a single catchment. It is possible from the data presented that Aboriginal burning actually began about a century before megafauna started to noticeably decline, and that final extinction of megafauna and decline in charcoal production occurred simultaneously. In any case, a century of sediment deposition was represented by less than two centimetres of soil depth, and the modelled age of sediments had error bounds of a millennium. Attempting to show that increased charcoal lagged behind reduced sporulation by only a century is like trying to split firewood with a scalpel, or maybe like trying to split hairs with a blockbuster.

In 2015, Tim Cohen and colleagues further muddied the waters around Aboriginal extinction of the megafauna by way of a hydrological study of the Lake Eyre Basin.[53] They claimed that the demise of *Genyornis* coincided with major climate change as evidenced by Lake Eyre drying out. Cohen's argument is unconvincing because the climate and the hydrology of the lake both changed far more abruptly, and the lake level dropped more dramatically, 20,000 years before the changes that he points to.[54] It also appears that Aboriginal use of the firestick was delivering more sediment into the Lake Eyre Basin despite decreasing streamflows as the bird became extinct.[55]

Bill Gammage's landmark work comprehensively documented historical evidence that the pre-European environment was maintained by Aboriginal burning.[56] He was awarded the Prime Minister's Literary Prize for Australian History as well as similar awards by the Premiers of Queensland and Victoria. One would have hoped that ecologists could no longer disparage Australia's fire history with impunity. Unfortunately, many are still publishing scientific articles portraying post-European scrubs and unnaturally dense stands of mallee or cypress or river red gum (Chapter 11) as if they are natural ecosystems requiring protection from human 'disturbance'. Considered objectively using the scientific method pre-history and history are the keys to understanding Australian ecology.

Endnotes

1 Gibbons *et al.* 2008.

2 Bowman 1998, ABC 1998.

3 Colloff 2014.

4 Pyne 1991.

5 Gammage 2011.

6 Jurskis and Underwood 2013.

7 Tench 1793.

8 Flinders 1814.

9 Oxley 1820.

10 Sturt 1833.

11 Mitchell 1839, 1848.

12 Darwin 1845.

13 Wallis 1878, Manton 1895.

14 Curr 1883.

15 Howitt 1891.

16 Jones 1969.

17 Croft *et al.* 1997.

18 Jurskis 2002.

19 Keith and Henderson 2002.

20 Henderson and Keith 2002.

21 Keith 2004.

22 Keith *et al.* 2002.

23 Jurskis and Underwood 2013.

24 Horton 1982.

25 Bowman 2003.

26 Bradstock *et al.* 2012, pp. 307-8.

27 Clarke *et al.* 2010, Haslem *et al.* 2011, Avitabile *et al.* 2013.

28 Clarke *et al.* 2010.

29 Haslem *et al.* 2011.

30 Mitchell 1839.

31 op. cit.

32 Clarke *et al.* 2010.

33 Avitabile *et al.* 2013.

34 Noble 1997.

35 Lunt 2014.

36 Flannery 1994.

37 MacPherson 1886, Parris 1948, Martin 1985, Low 2003.

38 Singh *et al.* 1981, Kershaw *et al.* 2002.

39 op. cit.

40 Mooney *et al.* 2011.

41 Flood 2006.

42 Mooney *et al.* 2012, Pyne 1991.

43 Mooney *et al.* 2011.

44 Burrows *et al.* 1995, Jurskis *et al.* 2006, Mooney *et al.* 2011.

45 Kershaw *et al.* 2002.

46 Singh *et al.* 1981.

47 Miller *et al.* 2005.

48 Prideaux *et al.* 2010.

49 Flannery 2012.

50 Rule *et al.* 2012.

51 Ibid., Figs. 1, 2B.

52 Ibid.

53 Cohen *et al.* 2015.

54 Ibid., Fig. 2A.

55 Ibid., Fig. 2C.

56 op. cit.

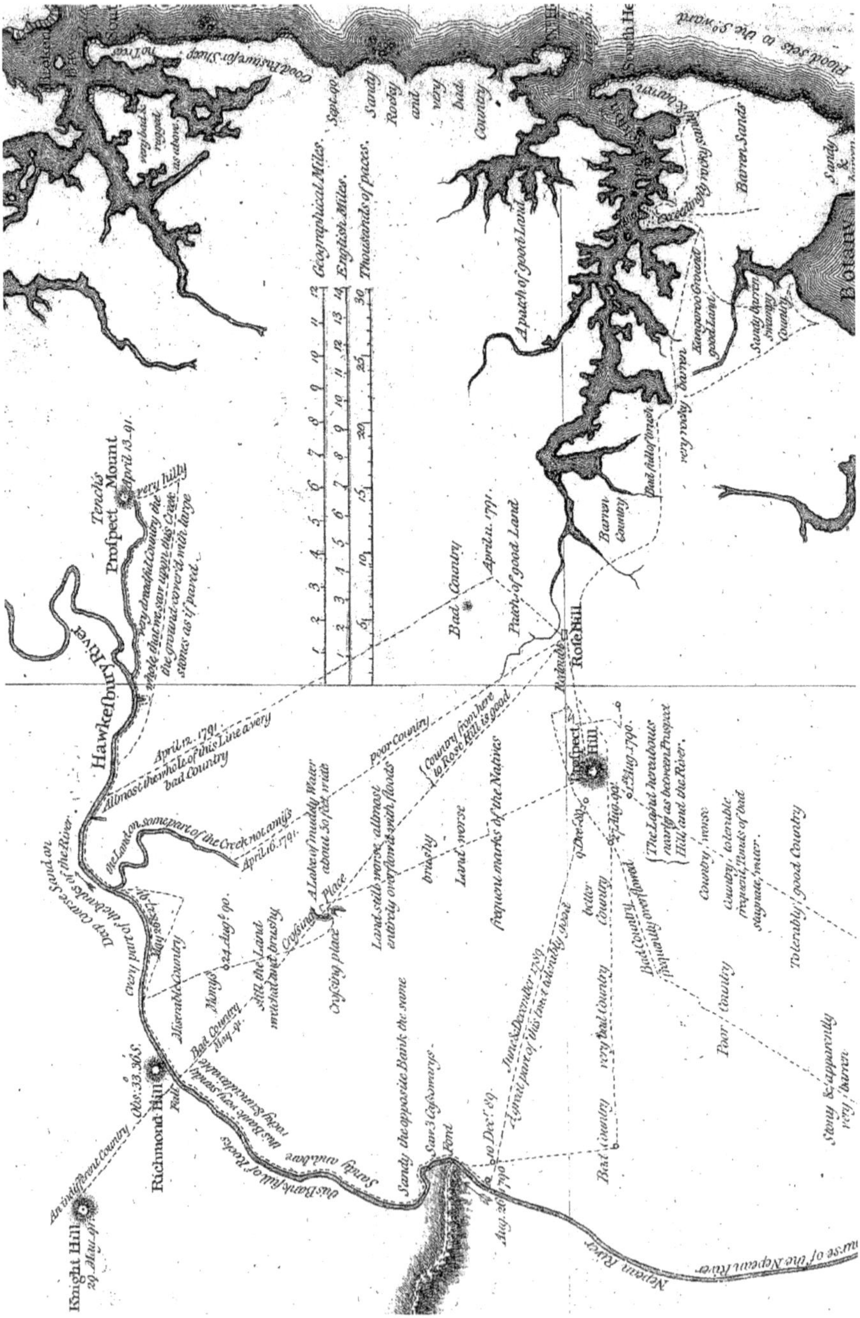

Fig. 2 The country between Botany Bay, Broken Bay and the Hawkesbury River. Watkin Tench 1793. (Courtesy State Library of NSW)

3

HOW THINGS WERE

To Everything (Turn, Turn, Turn)
There is a season (Turn, Turn, Turn)
And a time to every purpose, under heaven
Book of Ecclesiastes and Pete Seeger

The first Australians dramatically changed their environment and, over millennia, created a new, entirely predictable pattern of life. Climate, soils, topography and the firestick determined where plants and animals lived. The pattern was dynamic as a result of climatic variability, with habitats expanding and contracting according to variations at a range of time scales. European explorers and naturalists observed and explained the basic patterns of geology, topography, soils, human fire, flora and fauna, and they recognised climatic variability.

Palaeological history has been challenged by naturists such as Horton, and Benson and Redpath,[1] with perplexing arguments that Aboriginal arrival had little impact on the Australian environment. They grossly exaggerate the impacts of European settlers' axes and ploughs so as to promote a wilderness philosophy. To avoid any counter-accusations, in this chapter I will try to let the explorers speak for themselves as they journey across southeastern Australia and, in Darwin's case, cross Bass Straight and the Great Australian Bight. This will set a baseline for the following chapters. Where I have been unable to resist editorialising, it will be transparent.

Watkin Tench arrived with the First Fleet in 1788 and explored the area around Port Jackson (Sydney Harbour) over four years before returning to England. He published a fairly comprehensive natural history in 1793 as part of his account of European settlement.[2] Earlier, in 1790, Surgeon General John White had published a Journal with 65 plates of flora, fauna and Aboriginal implements.[3] Tench described the country between Broken Bay, Botany Bay and the Hawkesbury River (Fig. 2). He noted the lack of permanent streams, the flood risk on the Hawkesbury, and the general poverty of the soil.

There was some land suitable for cultivation to the west of Rose Hill (Parramatta) but other than that "the whole space of intervening country yet explored, (except a narrow strip called kanguroo ground) in both directions, is so bad as to preclude cultivation". The "kanguroo ground" lay between what are now Canterbury Park and Wentworth Park, and there were also small patches of good land immediately northeast of Parramatta and between Lane Cove and Willoughby. The map accompanying Tench's report shows many sandy, rocky, swampy and barren areas, but only two areas of brush, probably swamp oak (*Casuarina glauca*), tea tree (*Leptospermum, Kunzea*) and/or paperbark (*Melaleuca*), on boggy sites near the current Concord and Massey Park golf courses, and in the area that is now the suburb of Schofields.

> Of the soil, opinions have not differed widely. A spot eminently fruitful has never been discovered. That there are many spots cursed with everlasting and unconquerable sterility no one, who has seen the country, will deny. At the same time I am decidedly of the opinion, that many large tracts of land, between Rose Hill and the Hawkesbury, even now, are of a nature sufficiently favourable to produce moderate crops of whatever may be sown in them. And provided a sufficient number of cattle may be imported to afford manure for dressing the ground, no doubt can exist, that subsistence for a limited number of inhabitants, may be drawn from it. …
>
> The spontaneous productions of the soil, will soon be recounted. Every part of the country is a forest: Of the quality of the wood take the following instance. – The Supply wanted wood for a mast, and more than forty of the choicest young trees were cut down before as much wood as would make it could be procured: the trees being either rotten at the heart, or riven by the gum, which abounds in them. This gum runs not always in a longitudinal direction in the body of the tree, but is found in it in circles, like a scroll.

At the time, forest was a term used for widely spaced trees with grassy understorey[4] – woodland or open woodland in current terminology. Rotten hearts in young trees resulted from occlusion of dead sapwood as new wood grew over scars from repeated burning back of seedlings and saplings. Gum rings developed when saplings 'bled' after the cambium (live, inner bark) was scorched by fire. "The list of esculent vegetables, and wild fruits, is too contemptible to deserve notice, if the sweet tea, whose virtues have been already recorded and the common orchis root be excepted." White's journal included a Plate of sweet tea plant (*Smilax glyciphylla*) and reported that "The leaves have the taste of liquorice root accompanied with bitter. They are said to make a kind of tea, not unpleasant to the taste, and good for the scurvy".

> That species of palm-tree, which produces the mountain cabbage, is also found in most of the fresh water swamps, within six or seven miles of the coast; but is rarely seen further inland: even the banks of the Hawkesbury are unprovided with it. The inner part of the trunk of this tree was greedily eaten by our hogs, and formed their principal support. The grass, as has been remarked in former publications, does not overspread the land in a continuous sward, but arises in small detached tufts, growing every way about three inches apart, the intermediate space being bare; though the heads of the grass are often so luxuriant, as to hide all deficiency on the surface. The rare and beautiful flowering shrubs which abound in every part, deserve the highest admiration and panegyric.
>
> … The different temperatures of Rose Hill and Sydney, in winter, though only twelve miles apart, afford, however, curious matter of speculation. Of a well attested instance of ice being seen, at the latter place, I never heard. At the former place, its production is common, and once a few flakes of snow fell. The difference can be accounted for, only by supposing that the woods stop the warm vapours of the sea from reaching Rose Hill, which is at the distance of sixteen miles inland; whereas Sydney is but four. Again the heats of summer are more violent at the former place, than at the latter, and the variations incomparably quicker. … To convey an idea of the climate, in summer, I shall transcribe, from

my meteorological journal, accounts of two particular days, which
were the hottest we ever suffered under at Sydney.

December 27ᵗʰ 1790. Wind NNW; it felt like the blast of a heated
oven, and in proportion as it increased, the heat was found to be
more intense, the sky hazy, the sun gleaming through at intervals.

At 9 AM	85⁰
At noon	104
Half past twelve	107 1/2
From one P.M. until 20 minutes past two	108 1/2
At 20 minutes past two	109
At sunset	89
At 11 P.M.	78 1/2 …

This day's maximum of 109^0 F was equivalent to 43^0 C. By coincidence,
this was the temperature two centuries later at Death Valley, California
– the hottest known place on earth – at midday on 10ᵗʰ July 2013, the
Centenary of the day of the highest ever recorded temperature on earth
$(134^0$ F or 57^0 C).[5]

> But even this heat was judged to be far exceeded in the latter end
> of the following February, when the north-west wind again set in,
> and blew with a great violence for three days. At Sydney, it fell short
> by one degree of what I have just recorded: but at Rose Hill, it was
> allowed, by every person, to surpass all that they had before felt,
> either there, or in any other part of the world. Unluckily they had
> no thermometer to ascertain its precise height. It must, however,
> have been intense, from the effects it produced. An immense flight
> of bats, driven before the wind, covered all the trees around the
> settlement, whence they every moment dropped dead, or in a dying
> state, unable longer to endure the burning state of the atmosphere.'
> Nor did the *perroquettes*, though tropical birds, bear it better; the
> ground was strewed with them in the same condition as the bats.

Imagine the mileage that global warming enthusiasts such as Tim
Flannery and Will Steffen would make of such an event were it to happen
in Parramatta Park today.

Were I asked the cause of this intolerable heat, I should not hesitate to pronounce, that it was occasioned by the wind blowing over immense desarts, which, I doubt not, exist in a north-west direction from Port Jackson, *and not from fires kindled by the natives*. ...

My other remarks on the climate will be short; it is changeable beyond any other I ever heard of; ...

The leading animal production is well known to be the kanguroo. ...

Hitherto I have spoken only of the large, or grey kanguroo, to which the natives give the name of Pat-ag-a-ran. But there are (besides the kanguroo-rat) two other sorts. One of them we called the red kanguroo, from the colour of its fur, which is like that of a hare, and sometimes is mingled with a large portion of black: the natives call it Bag-a-ray [swamp wallaby]. It rarely attains to more than forty pounds weight. The third sort is very rare, and in the formation of its head, resembles the opossum [probably the brushtailed rock wallaby]. – The kanguroo-rat [rufous bettong] is a small animal, never reaching, at its utmost growth, more than fourteen or fifteen pounds, and its usual size is not above seven or eight pounds. It joins to the head and bristles of a rat, the leading distinctions of a kanguroo, by running, when pursued, on its hind legs only, and the female having a pouch. Unlike the kanguroo, which appears to have no fixed place of residence, this little animal constructs for itself a nest of grass, on the ground, of a circular figure, about ten inches in diameter, with a hole in one side, for the creature to enter at; the inside being lined with a finer sort of grass, very soft and downy. But its manner of carrying the materials with which it builds the nest, is the greatest curiosity: – by entwining its tail (which, like all of the kanguroo tribe, is long, flexible, and muscular) around whatever it wants to remove, and thus dragging along the load behind it. This animal is good to eat; but whether it be more prolific at birth than a kanguroo, I know not.

The Indians sometimes kill the kanguroo; but their greatest destroyer is the wild dog, who feeds on them. Immediately on hearing, or seeing, this formidable enemy, the kanguroo flies to the thickest cover, in which, if he can involve himself, he generally escapes. In running to the cover, they always, if possible, keep in paths of their own forming, to avoid the high grass, and stumps of trees, which might be sticking up among it, to wound them, and impede their course.

Our methods of killing them were but two; either we shot them, or hunted them with greyhounds; we were never able to ensnare them. Those sportsmen who relied on the gun, seldom met with success, unless they slept near covers, into which the kanguroos were wont to retire at night, and watched with great caution and vigilance when the game, in the morning, sallied forth to feed. ...

Other quadrupeds, beside the wild dog, consist only of the flying squirrel, of three kinds of opposums, and some minute animals, usually marked by the distinction which so peculiarly characterises the opossum tribe. ...

... A squirrel-trap ... When the Indians in their hunting parties set fire to the surrounding country (which is a very common custom) the squirrels, opposums, and other animals who live in trees, flee for refuge into these holes, whence they are easily dislodged and taken.

Tench discussed the "cassowary or emu" at some length, and mentioned fourteen sorts of parrots, numerous hawks and quails, a snipe, ducks, geese and "other aquatic birds" as well as some "very beautiful" smaller birds including the "coach-whip".

... The country, I am of opinion, would abound with birds, did not the natives, by perpetually setting fire to the grass and bushes, destroy the greater part of the nests; a cause which also contributes to render small quadrupeds scarce: they are besides ravenously fond of eggs, and eat them wherever they find them.

Tench also mentioned that there were more than ten kinds of saltwater fish and two or three kinds of snakes.

... the natives shew, on all occasions, the utmost horror of the snake, and will not eat it, although they esteem lizards, guanas, and many other reptiles, delicious fare ...

Summer here, as in all other countries, brings with it a long list of insects. In the neighbourhoods of rivers, and morasses, musquitoes and sand flies are never wanting at any season; but at Sydney, they are seldom numerous, or troublesome. The most nauseous, and destructive of all the insects, is a fly, which blows not eggs, but large living maggots; and if the body of the fly be opened it is found full of them.

(Commonly known as flesh flies, these have been the subject of much discussion and debate by taxonomic entomologists in Australia because identification of species[6] often relies upon microscopic examination of the male genitalia.)

"Of ants, there are several sorts, one of which bites very severely: the white ant is sometimes seen. Spiders are large and numerous. Their webs are not only the strongest, but the finest, and most silky, I ever felt".

Tench never saw or heard of a koala because they were rare under Aboriginal management. John Price, a young servant of Governor Hunter, first recorded the koala, the wombat and the lyrebird on an exploration near what is now known as Pheasants Nest (for the lyrebird) on the 26th January 1798,[7] exactly ten years after first settlement.

Tench summarised his anthropological observations in seventeen pages including many references to use of fire. For example:

Their method of procuring fire is this: they take a reed, and shave one side of the surface flat; in this they make a small incision to reach the pith, and introducing a stick, purposely blunted at the end, into it, turn it round between the hands (as chocolate is milled) as swiftly as possible, until flame be produced. As this operation is not only laborious, but the effect tedious, they frequently relieve each other of the exercise. And to avoid being often reduced to the necessity of putting it in practice, they always, if possible, carry a lighted stick with them, whether in their canoes, or moving from place to place on land.

The new colony was hemmed in by the Blue Mountains. Tench wrote "from Knight Hill (Scott Trig at Panorama Point, Kurrajong Heights[8]), and Mount Twiss (on Dawes Ridge north of Woodford[9]), the limits which terminate our researches, nothing but precipices, wilds, and desarts are to be seen".

Blaxland, Lawson and Wentworth led the first party of Europeans across the Blue Mountains in 1813. Blaxland (using the third person) described the difficulties they encountered ascending the precipitous sandstone country:[10]

> The land was covered with scrubby brush-wood, very thick in places, with some trees of ordinary timber, which much incommoded the horses. The greater part of the way they had deep rocky gullies on either side of their track, and the ridge they followed was very crooked and intricate. In the evening they decamped at the head of a deep gully, which they had to descend for water; they found but just enough for the night, contained in a hole in the rock, near which they met with a kanguroo, who had just been killed by an eagle. A small patch of grass supplied the horses for the night. ...
>
> They found it impossible to travel through the brush before the dew was off, and could not, therefore, proceed at an earlier hour in the morning than nine. After travelling about a mile on the third day, in a west and north-west direction, they arrived at a large tract of forest land [now Springwood Ridge], rather hilly, the grass and timber tolerably good, extending, as they imagine, nearly to Grose Head, in the same direction nearly as the river. They computed it at two thousand acres. Here they found a track marked by a European, by cutting the bark of the trees. Several native huts presented themselves at different places. They had not proceeded above two miles, when they found themselves stopped by a brushwood much thicker than that which had hitherto met with. This induced them to alter their course, and to endeavor to find another passage to the westward; but every ridge which they explored proved to terminate in a deep rocky precipice; and they had no alternative to return to the thick brushwood, which appeared to be the main ridge, with the determination to cut a way through for the horses the next day...
>
> The party encamped in the forest tract, with plenty of good grass and water.

Lawson mentioned[11] that the "forest land" contained "great Quantitys of Indigo [*Indigofera australis*] growing much such land [sic] as Lane Cove [one of the kanguroo grounds mentioned by Tench] – found several camps of native huts".

The thick brushwood was tall shrubland in modern terminology. Lawson reported "Cut a Road about five miles through a thick brush this is a very poor Rocky and Sandy Country I ever saw [sic] with great quantitys of Honey Suckle growing and the gullys extremely deep". Blaxland also identified honeysuckle or *Banksia* as a dominant component whilst Wentworth[12] added mention of "tea tree", "pear tree" (*Xylomelum*), "currant bushes" (*Coprosma*), "peppermint" (*Eucalyptus piperita*), "and in general the same variety of small flowering scrubs which grow between Sydney and the South Head". (White's Plate of *E. piperita* depicts seed capsules of a stringybark, *E. globoidea*, and the botanical description refers to "the ripe capsules of this tree (although not attached to the specimens of the leaves)".)

Wentworth described the shrubby woodlands and eucalypt forests thus:

> … dwarf gums, stringybarks and blood trees [*Corymbia*] all of which are generally withered on the tops –The gullies however for the first 15 miles produce a tree very much resembling the Mountain Ash which grows very luxuriantly and is most probably a very valuable Wood – For the last 10 miles also which we travelled on this range the stringy barks and the gums were of more stately growth altho there is no visible alteration in the Soil – This change arises most probably from some alteration in the Substratum …

Wentworth's reference to mountain ash is intriguing. It is obviously neither the Victorian eucalypt that now has this common name, nor Blue Mountains ash, *E. oreades*, which doesn't grow in the lower eastern parts. He was most likely referring to one of two species of *Polyscias* – rainforest trees which grow in deep sheltered gullies and are self-pruning, with a straight trunk, though usually slender with a maximum diameter of about 50 centimetres. The pinnate leaves of these species have some resemblance to European mountain ash or *Sorbus*, and their straight, symmetrical trunks, clear of branches or stubs are suggestive of straight grained knot-free timber. The timber is clean and straight grained but it is not strong or durable.

The three explorers described grasslands/sedgelands, swamps and low heaths with their characteristic substrates in more or less detail. Wentworth observed that:

> It is the Moisture which the Mountains extract from the clouds which gives rise to the Innumerable small streams which every where pervade the Mountains and eventually furnish all the rivers on this side of the Country with a constant supply of water.

After descending to the Central Tablelands at what is now Mount York, Blaxland wrote:

> They reached the foot at nine o'clock a.m., and proceeded two miles north-north-west, mostly through open meadow land, clear of trees, the grass from two to three feet high. They encamped on the bank of a fine stream of water. The natives, as observed by the smoke of their fires, moved before them as yesterday. The dogs killed a kanguroo, which was very acceptable, as the party had lived on salt meat since they caught the last. The timber seen this day appeared rotten and unfit for building.
>
> Sunday, the 30th [of May], they rested in their encampment. One of the party shot a kanguroo with his rifle, at a great distance across a wide valley. The climate here was found very much colder than that of the mountain, or of the settlements on the east side, where no signs of frost had made its appearance when the party set out. During the night the ground was covered with a thick frost, and a leg of the kanguroo was quite frozen. From the dead and brown appearance of the grass it was evident that the weather had been severe for some time past. We were all much surprised at this degree of cold and frost in the latitude of about 34 degrees. The track of the emu was noticed at several places near the camp.
>
> On the Monday they proceeded about six miles, south-west and west, through forest land, remarkably well watered, and several open meadows, clear of trees, and covered with high good grass. They crossed two fine streams of water. Traces of the natives

presented themselves in the fires they had left the day before, and
in the flowers of the honeysuckle tree scattered around, which had
supplied them with food. These flowers, which are shaped like a
bottle-brush, are very full of honey. The natives on this side of the
mountains appear to have no huts like those on the eastern side,
nor do they strip the bark or climb the trees. From the shavings
and pieces of sharp stone which they had left, it was evident that
they had been busy sharpening their spears.

The Aboriginal people in this area presumably sheltered under
overhanging rock in the sandstone escarpments at the heads of the
valleys, and mostly speared game on the ground rather than extracting
possums from trees.

Wentworth contrasted the fauna of the grassy woodlands against that
of the shrubby sandstone country:

> During the whole of our excursion we saw but very few Kangaroos
> and those mostly in the forest land – We saw however a few
> Kangaroo Rats bandicoots and brush Kangaroo on the Mountain
> but I do not think that either the Squirrel or Opossum frequent
> the Mountain at all.

Later in the same year, Assistant Surveyor George Evans extended the
explorations from the eastern edge of the central tablelands to the western
slopes, as summarised by Secretary of the Colony, J.T. Campbell:[13]

> … he arrived at the termination of the tour lately made by Messrs.
> G. Blaxland, W. C. Wentworth and Lieutenant Lawson. Continuing
> in the western direction prescribed in his instructions, for the
> course of twenty-one days from this station, Mr. Evans then found
> it necessary to return; and on the 8th of January he arrived back at
> Emu Island, after an excursion of seven complete weeks. During
> the course of this tour Mr. Evans passed over several plains of great
> extent, interspersed with hills and valleys, abounding in the richest
> soil, and with various streams of water and chains of ponds.

Evans turned back after he reached more rugged country on the
Macquarie River northwest of the Bathurst Plains:[14]

I made up my mind to return in the Morning, seeing no hopes
of approaching the end of the high Range of Hills; I would most
willingly proceed farther, but the Horses backs being so bad; nor
can you have an Idea of the situation we are in with respect to our
feet; with patching and mending we may manage to reach home. I
am now 98½ measured Miles from the limitation of Mr. Blaxland's
excurtion; most part of the distance is through a finer Country than
I can describe, not being able for want of Language to dwell on the
subject, or explain its real and good appearance with Pen and Ink,
… I have succeeded in passing over a Beautiful Country, and make
no doubt but that to the Westward of these hills there may be a
part equal to it; also beg leave to say I shall be happy and ready to
go on at any future time to attempt a Journey to the Western coast,
which I think this river [The Macquarie] leads to; it is a rapid Stream
in the Winter Seasons, is of great width there being two Banks. The
Hollow, which I imagine from the hills to be its course, bears North
of West. I conceive it strange we have not fell in with the Natives;
they are near about us as we find late traces of them; I think they are
watching us, but are afraid and keep at some distance. …

Fine weather very warm; halted at the commencement of Bathurst
Plains early, as I was desirous to examine this part; I ascended Mount
Pleasant, the West end led me on a Ridge of Beautiful hills, along
which I travelled about 3 Miles, a small stream of Water forming
ponds run at their foot; I was gratified with a pleasing sight of an
open Country to the S.W. of them; at the space of 7 or 8 Miles I could
discern the Course of a River winding to the West [now Belabula
River, a tributary of the Lachlan]; I saw three or four large Plains;
the first of them I was on, the Chain of Ponds before mentioned
running through it; I feel much regret I am not able to Travel a week
or more in that direction; I imagine the flat open Country extends
30 or 40 Miles; at the termination I can only discern one Mountain
Quite Pale with three Peaks; … I cannot speak too much of the
Country, the increase of Stock for some 100 Years cannot overrun
it; the Grass is so good and intermixed with variety of herbs. Emu's
and Geese are numerous, … we counted 41 Emu's this day; …
Returning we saw smoke on the North side of the River, at Sun sett
as we were fishing I saw some Natives coming down the Plain; they

did not see us untill we surprised them: there was only two Women
and four Children, the poor Creatures trembled and fell down with
fright; I think they were coming for Water; I gave them what Fish
we had, some fish Hooks, Twine and a Tomahawk, they appeared
glad to get from us;

Mr Evans' opinion of the land's carrying capacity, based on a very
good season, was overly optimistic, and official exploration barely kept
ahead of settlement. In proposing a journey down the Macquarie River
to the west coast, Evans' gave no estimate of distance. However Colonial
Secretary Campbell seriously underestimated the likely distance: "This
river is supposed to empty itself into the ocean, on the western side of
New South Wales, at a distance of from two to three hundred miles from
the termination of the tour.". This is rather perplexing because Flinders
had circumnavigated Australia a decade earlier and had found no large
river on the western or southern coasts. The mouth of the Murray was
probably closed at that time. Flinders recognised marine sediments in the
cliffs of the Great Australian Bight and speculated that they "may even
be a narrow barrier between an interior, and the exterior sea".[15] But these
cliffs are 900 miles west of the termination of Evans' tour.

On his return journey, during the first week of January (midsummer)
in 1814, Evans observed Aboriginal burning between what is now
Katoomba, and the Nepean River:

> The Mountains have been fired; had we been on them we could
> not have escaped; the Flames rage with violence through thick
> underwood, which they are covered with. Bad travelling the stick
> of the Bushes here are worse than if their leaves had not been
> consumed; …
>
> The Mountains are as yesterday; fired in all directions; at 11
> o'clock I was upon the high hill; all objects Eastward are obscured
> by thick smoke; We stopped where there was feed for the Horses
> and Water. …
>
> Still a thick Brush; the leaves of it are burnt. The weather is
> disagreeably warm and boisterous, which has been the case these
> last 3 days. I halted on the top of a Mountain … with the hope

that I shall be able to see the Hawkesbury from it, should it turn out a clear Morning. There is water and sufficient feed for our Horses. …

… The Ridges continue as usual until the latter part of my journey which is Forest land for ½ a Mile; the timber on it is chiefly lofty stringy Bark and Oaks: there are small patches of Grass left that the fire missed. …

… The Forest land continues a Mile farther; afterwards the brushy Ridge commences again, the thickest of it is consumed, which I consider fortunate, had it not I should be obliged to have given off measuring; at the end of today's Journey is a Lagoon of good Water, with tolerable grass round the edge of it. …

¾ of a Mile terminates the brush, the ridges then produce good sheep feed for 1½ Mile, when there is a gully which is the south side of *Emu Island*.

It is notable that moderate and high-intensity fires burning in thick brush, under warm and stormy conditions, did not develop into megafires, but died down or stopped in "forest land".

In 1815, Evans explored the upper Lachlan River:

The River I can distinctly discover to continue near due West, and rest confident that, when it is full, Boats may go down it in safety; my meaning of being full is its general height in moderate Seasons, which the banks shew, about Five feet above the presant level; it would then carry Boats over Trees and Narrows that now obstruct the Passage; no doubt the Stream connects with Macquarie or some other River further West; the Channel then sure is of great magnitude; I should think so to carry off the body of Water that must in time of Floods cover these very extensive flats. …

Left the River, which I have now called *"The River Lachlan"*.[16]

Late in 1815, Secretary of State for The Colonies, Earl Henry Bathurst, learnt of the crossing of the Blue Mountains, and the discovery of the Macquarie. Governor Lachlan Macquarie received a dispatch from Bathurst early in 1817 containing congratulations and instructions to explore the Macquarie by boat:[17]

The Discovery of a Passage on the Blue Mountains, and of an extensive Country beyond them, possessing so many Advantages in point of Soil and Climate, is an Event in the History of New So. Wales not more important on Account of its immediate Consequences than for the Effect, which it must ultimately produce in the Nature and Value of the Colony. ...

... one fact in Mr. Evans' Journal, the Discovery of the Macquarie River, excited general Surprise; for after the fruitless Attempt made by Captain Flinders to discover, on the West Coast of New Holland, the Embouchure of any considerable River, I was but little prepared to expect that a River should have so soon been discovered in the Interior of that Continent flowing for so considerable a distance to the Westward. In a Geographical point of View, it is most desirable to ascertain the Course of this River; ... and it therefore appears adviseable that a small Vessel should be either constructed on the Banks of the Macquarie River itself, or conveyed thither in frame, and set up on that part of the River deemed most convenient for the Commencement of the Navigation.

Bathurst was clearly aware that Flinders had found no substantial river mouth on Australia's southern or western coasts. Somehow, Campbell had formed the opinion that the mouth of the Macquarie was only three hundred miles west of the Bathurst Plains. As it turned out, Evans' report of the potentially navigable Lachlan River flowing due west induced the Governor to change the plan. Surveyor-General John Oxley was instructed to descend the Lachlan River:[18]

On the twenty-fourth of March [1817] I received the instructions of His Excellency the Governor to take charge of the expedition which had been fitted out for the purpose of ascertaining the course of the Lachlan River, and generally to prosecute the examination of the western interior of New South Wales.

Oxley was blocked by extensive morasses on the floodplain of the Lachlan below what is now Booligal. "I was forced to come to the conclusion, that the interior of this vast country is a marsh and uninhabitable. How near these marshes may approach the south-western coast, I know not."

He travelled overland to the Macquarie, returned to Bathurst, and reported that:

> Nothing can afford a stronger contrast than the two rivers, Lachlan and Macquarie; different in their habit, their appearance, and the sources from which they derive their waters, but above all differing in the country bordering on them; the one constantly receiving great accession of water from four streams, and as liberally rendering fertile a great extent of country; whilst the other from its source to its termination, is constantly diffusing and extenuating the waters it originally receives over low and barren deserts creating only wet flats and uninhabitable morasses, and during its protracted and sinuous course is never indebted to a single tributary stream.

Oxley later added:

> it is the intention of His Majesty's Government to follow the course of the Macquarie River, and it is sanguinely expected that the result of the contemplated expedition will be such as to leave no longer in doubt the true character of the country comprising the interior of this vast island. It would be as presumptuous as useless to speculate on the probable termination of the Macquarie River, when a few months will (it is to be hoped) decide the long disputed point, whether Australia, with a surface nearly as extensive as Europe, is, from its geological formation, destitute of rivers, either terminating in interior seas, or having their estuaries on the coast.

But his hopes were dashed when he reached the Macquarie Marshes where the character of the river is not much different to the lower Lachlan. Flinders' earlier speculation about an inland sea and Campbell's inexplicable estimate of its distance to the west were at this time seemingly validated. However, Military Secretary Charles Sturt navigated the Murrumbidgee and Murray Rivers to the mouth at Lake Alexandrina on the south coast in 1829/30.[19] Sturt contributed to paleoecology as well as historical ecology. His observations of marine fossils in the banks of the Murray proved that the fabulous 'inland sea' was an anachronism rather than a fiction. When he reached the point in the Lake where the Murray's right bank terminates in a promontory Sturt wrote:

Thus far, the waters of the lake had continued sweet; but on filling a can when we were abreast of this point, it was found that they were quite unpalateable, to say the least of them. The transition from fresh to salt water was almost immediate.

The tide in the river at the top of the lake was measured by Sturt at "eight inches" (200 millimetres). When he attempted to navigate the entrance channel at the bottom of the lake "fifty-three miles away: just as we were about to push from the shore, a seal rose close to the boat, which we all regarded as a favourable omen. We were, however, shortly stopped by shoals". The mouth of the Murray was most probably closed when Flinders charted our southern coastline in 1802, but it was obviously open in 1830. Sturt noted on his return journey to the Murrumbidgee that places he had formerly condemned as barren, were green and luxuriant.

Surveyor-General Major Thomas Mitchell F.G.S. & M.R.G.S. "traced out the new line of descent from the Blue Mountains to the interior country, by the pass which I then named Mount Victoria" in 1830. He explored most of the Murray-Darling Basin and some areas beyond in four expeditions between 1831 and 1846.[20] When Mitchell reached the Lachlan River in autumn of 1836 he noted:

> But its waters were gone, except in a few small ponds, in the very deepest parts of its bed. Such was now the state of that river, down which my predecessor's boats had floated. I had, during the last winter, drawn my whale boats 1,600 miles [2,600 kilometres] over land without finding a river, where I could use them ; whereas, Mr Oxley had twice retired by nearly the same routes, and in the same season of the year, from supposed inland seas !

Upon reaching Cudjallagong (now Lake Cargelligo) which had been described by Oxley as a "noble lake", Mitchell remarked thus upon the extreme variability of Australia's climate:

> On its northern margin and a good way within the former boundary of the lake, stood dead trees of a full grown size, which had been apparently killed by too much water, plainly shewing, like the trees similarly situated in Lake George and Lake Bathurst, to what long

periods, the extremes of drought and moisture have extended, and may again extend, in this singular country.[21]

During *The Voyage of the Beagle*,[22] Charles Darwin descended Mount Victoria pass in January 1836:

> We now entered upon a country less elevated by nearly a thousand feet, and consisting of granite. With the change of rock, the vegetation improved, the trees were both finer and stood further apart; and the pasture between them was a little greener and more plentiful. At Hassan's Walls, I left the high road, and made a short detour to a farm called Walerawang; to the superintendent of which I had a letter of introduction from the owner in Sydney. Mr. Browne had the kindness to ask me to stay the ensuing day, which I had much pleasure in doing. This place offers an example of one of the large farming, or rather sheep-grazing establishments of the colony. Cattle and horses are, however, in this case rather more numerous than usual, owing to some of the valleys being swampy and producing a coarser pasture. Two or three flat pieces of ground near the house are cleared and cultivated with corn ...
>
> ... Early on the next morning Mr. Archer, the joint Superintendent, had the kindness to take me out kangaroo-hunting. We continued riding the greater part of the day, but had very bad sport, not seeing a kangaroo, or even a wild dog. The greyhounds pursued a kangaroo rat into a hollow tree, out of which we dragged it: it is an animal as large as a rabbit, but with the figure of a kangaroo. A few years since this country abounded with wild animals; but now the emu is banished to a long distance, and the kangaroo is become scarce; to both the English greyhound has been highly destructive. It may be long before these animals are altogether exterminated; but their doom is fixed. The aborigines are always anxious to borrow the dogs from the farm-houses; the use of them, the offal when an animal is killed, and some milk from the cows, are the peace-offerings of the settlers who push farther and farther towards the interior. ...

Rather than doom, kangaroos experienced a boom a few decades later. Curr wrote:[23]

After the aborigines and the wild dog (originally the meat consum-
ers of the continent) had been got rid of, the animals on which
they preyed increased enormously in many localities. As regards
the kangaroo, indeed, matters became very serious. On an average
these animals, it is thought, consume as much grass as a sheep,
and where a few score had originally existed there soon came to
be a thousand ; so that, in some places, they threatened to jostle
the sheep and his master out of the land ; and, in consequence,
energetic and costly steps had to be taken to reduce their numbers. As an instance of the proportions to which the kangaroo
plague attained in Victoria–not to speak of Queensland, where
the evil has become much more serious–I may mention a property
of some 60,000 or 80,000 acres, on which ten thousand of these
animals were killed and skinned annually for six consecutive years,
their numbers still remaining very formidable in the locality. In
other places, by means of long lines of fencing and high yards,
specially erected for the purpose, as many as two thousand have
sometimes been destroyed in a single day.

At the "Conclusion" of his *Recollections* … Curr undermined his own
argument that Aborigines and dingoes had formerly kept kangaroo
numbers under control, by boasting of his success as a squatter:

> I think it will be admitted that the undertaking was brought to a
> successful issue, especially when it is remembered that, in addition
> to the thirty thousand sheep which I left on the ground in 1851, the
> run secured was of first-class quality as any squatting country, and
> capable, with the help of a few tanks, of depasturing a hundred
> thousand sheep. Indeed, I believe I was successful as any sheep-
> farmer of those times.

The irruption of kangaroos that Curr attributed to release from
hunting pressure, was actually due to the provision of permanent water.
A run of good seasons and the construction of many tanks, dams and
bores during the latter half of the nineteenth century allowed both
kangaroos and graziers to flourish.[24] If pastoralists, each killing thousands
of 'roos each year didn't make a dent in the population, it's hardly likely
that Aborigines and dingoes had previously kept them in check.

The temporary scarcity of kangaroos and emus that had been attributed variously to hunting by Aborigines and dogs was a result of drought. Indeed Tench had remarked upon the kangaroo's lack of fecundity during the first drought suffered by European settlers: "Prolific it cannot be termed, bringing forth only one at birth, which the dam carries in her pouch wherever she goes." Although offspring emerge singly, female kangaroos are normally continuously pregnant with a quiescent embryo resuming development as soon as its older sibling leaves the pouch. There is some overlap in suckling, with the weaner faithfully returning to its original teat and the dam producing quite different milk from the separate teats for the different aged offspring. Poor nutrition during droughts causes a suspension of the oestrous cycle so that a female does not mate after giving birth.[25]

Resuming our excursion with Darwin at "Walerawang":

Although having poor sport, we enjoyed a pleasant ride. The woodland is generally so open that a person on horseback can gallop through it. It is traversed by a few flat-bottomed valleys, which are green and free from trees: in such spots the scenery was pretty like that of a park. In the whole country I scarcely saw a place without the marks of a fire; whether these had been more or less recent – whether the stumps were more or less black, was the greatest change which varied the uniformity, so wearisome to the traveller's eye. In these woods there are not many birds; I saw, however, some large flocks of the white cockatoo feeding in a cornfield, and a few most beautiful parrots; crows, like our jackdaws were not uncommon, and another bird something like the magpie. In the dusk of the evening I took a stroll along a chain of ponds, which in this dry country represented the course of a river, and had the good fortune to see several of the famous Ornithorhyncus paradoxus. They were diving and playing about the surface of the water, but showed so little of their bodies, that they might easily have been mistaken for water-rats. Mr. Browne shot one: certainly it is a most extraordinary animal; a stuffed specimen does not at all give a good idea of the appearance of the head and beak when fresh; the latter becoming hard and contracted.

In *The Origin of Species*, platypus figured prominently amongst Darwin's arguments around missing links. For example:

> in fresh water we find some of the most anomalous forms now known in the world as the Ornithorhynchus and Lepidosiren [lungfish] which, like fossils, connect to a certain extent orders at present widely sundered in the natural state. ... It accords with the widely extended principle of specialisation, that the glands over a certain space of the sack should have become more highly developed than the remainder; and they would then have formed a breast, but at first without a nipple, as we see in the Ornithorhynchus, at the base of the mammalian series.[26]

Resuming *The Voyage of the Beagle*:

> ... A long day's ride to Bathurst. Before joining the highroad we followed a mere path through the forest; and the country with the exception of a few squatters' huts, was very solitary. We experienced this day the sirocco-like wind of Australia, which comes from the parched deserts of the interior. Clouds of dust were travelling in every direction; and the wind felt like it had passed over a fire. ... In the afternoon we came in view of the downs of Bathurst. These undulating but nearly smooth plains are very remarkable in this country, from being absolutely destitute of trees. They support only a thin dry pasture. ... The season, it must be owned, had been one of great drought, and the country did not wear a favourable aspect; although I understand it was incomparably worse two or three months before. ... The Macquarie figures in the map as a respectable river, and it is the largest of those draining this part of the water-shed; yet to my surprise I found it a mere chain of ponds, separated from each other by spaces almost dry. Generally a small stream is running; and sometimes there are high and impetuous floods. Scanty as the supply of water is throughout this district, it becomes still scantier further inland. ...

> The next day we passed through large tracts of country in flames, volumes of smoke sweeping across the road. Before noon we joined our former road and ascended Mount Victoria.

It is interesting that Darwin and his local guide travelled apparently unconcerned amongst extensive fires driven by strong winds, in the middle of summer, during a severe drought.

The Beagle also stayed ten days at Hobart Town where Darwin observed some moist eucalypt forest (see Chapter 12):

> The climate here is damper than in New South Wales, and hence the land is more fertile. Agriculture flourishes; … Another day I ascended Mount Wellington; I took with me a guide, for I failed in a first attempt, from the thickness of the wood. Our guide, however, was a stupid fellow, and conducted us to the southern and damp side of the mountain, where the vegetation was very luxuriant; and where the labour of the ascent, from the number of rotten trunks, was almost as great as on a mountain in Tierra del Fuego or in Chiloe. It cost us five and a half hours of hard climbing before we reached the summit. In many parts the Eucalypti grew to a great size, and composed a noble forest. In some of the dampest ravines, tree-ferns flourished in an extraordinary manner; I saw one which must have been at least twenty feet high to the base of the fronds, and was in girth exactly six feet. The fronds forming the most elegant parasols, produced a gloomy shade, like that of the first hour of the night. The summit of the mountain is broad and flat, and is composed of huge angular masses of naked greenstone. Its elevation is 3100 feet above the level of the sea. The day was splendidly clear, and we enjoyed a most extensive view; to the north, the country appeared a mass of wooded mountains, about the same height with that on which we were standing, and with an equally tame outline: to the south the broken land and water, forming many intricate bays, was mapped with clearness before us. After staying some hours on the summit, we found a better way to descend, but did not reach the Beagle till eight o'clock, after a severe day's work.

The summer day was clear and the view was extensive because there were no fires burning. Virtually all surviving Aborigines had been deported from Tasmania to Flinders Island by 1834.[27] Departing Tasmania, the Beagle proceeded to King George Sound where Darwin observed tall open shrublands and low woodlands:

February 7th. – The Beagle sailed from Tasmania, and, on the 6th of the ensuing month, reached King George's Sound, situated close to the S. W. corner of Australia. We stayed there eight days; and we did not during our voyage pass a more dull and uninteresting time. The country, viewed from an eminence, appears a woody plain, with here and there rounded and partly bare hills of granite protruding. One day I went out with a party, in hopes of seeing a kangaroo hunt, and walked over a good many miles of country. Everywhere we saw the soil sandy, and very poor; it supported either a coarse vegetation of thin, low brushwood and wiry grass, or a forest of stunted trees. The scenery resembled that of the high sandstone platform of the Blue Mountains; the Casuarina (a tree somewhat resembling a Scotch fir) is, however, here in greater number, and the Eucalyptus in rather less. In the open parts there were many grass-trees, – a plant which, in appearance, has some affinity with the palm; but, instead of being surmounted by a crown of noble fronds, it can boast merely a tuft of very coarse grass-like leaves. The general bright green colour of the brushwood and other plants, viewed from a distance, seemed to promise fertility. A single walk, however, was enough to dispel such an illusion; and he who thinks with me will never wish to walk again in so uninviting a country.

Darwin's observations that grasstrees had only a tuft of green leaves (without a skirt of dry leaves), and that the brushwood had bright green leaves are a sure indication that the area had been recently burnt by Aborigines with mild fire that burnt the skirts of grasstrees, and scorched the leaves of shrubs and small trees, producing a flush of new growth.

Early descriptions of the Australian landscape were often gloomy. Curr contrasted reports by Oxley and Sturt against those by Mitchell who could generally find something positive to say about the Australian people and landscape:

The explorers who first entered Riverina were Oxley and Sturt. Oxley's was what might be called a trial trip, and a very gallant one, and no old hand forgets that "Sturt's furthest" held the post of honour on our maps for nearly twenty years, and that he was a most energetic explorer. But though I value highly the labours of these leaders and their results, I may remark that the records of their travels produce the impression that they suffered from a great deal of despondency as they wended their way through the woods at the heads of their small parties. ... Transplanted from an island, continental features affected them unpleasantly. Forests which took weeks to traverse ; plains like the ocean, horizon unbounded ; the vast length of our rivers when compared to those of England, often flowing immense distances without change or tributary, now all but dry for hundreds of miles, at other times flooding the countries on their banks to the extent of inland seas, wearied them more in the contemplation than by travel.

... As for Mitchell, Leichardt, and other explorers, they had imbibed a good deal of the knowledge and ideas of persons born and brought up in the bush ; and, instead of suffering from despondent views, we find them, on the contrary, with a fresh paradise to report at the end of each exploration.[28]

Mitchell's enthusiastic contributions to diverse fields of science in Australia[29] seem now to be rather undervalued. *Three Expeditions into the Interior of Eastern Australia* and *Journal of an Expedition into the Interior of Tropical Australia* not only present valuable insights into Australian geomorphology, prehistory, climate, soils, ecology, anthropology, history and sociology but they also exemplify the value of observation and deduction. Education of young Australians would benefit greatly if these books were prescribed reading in high school curricula.

Of geomorphology Mitchell wrote:[30]

A yellow highly calcareous sandstone, apparently stratified, occurs near the banks of the Gwydir. Large rounded boulders of argillaceous limestone have been denuded in the bed of Glendon brook ; and an impure limestone is found in the neighbourhood of William's river, both belonging to the basin of the Hunter, and not much elevated above the sea. Calcareous tuff or grit, may be observed in various localities – and calcareous concretions abound in the blue clay of almost all the extensive plains – on both sides of the mountains.

A soft shelly limestone, most probably of recent origin, though slightly resembling some of the oolites of England, occurs extensively on the southern coast, between Cape Northumberland and Portland bay, where it forms the only rock with the exception of amygdaloidal trap.

Granite, or granitic compounds, are more or less apparent at or near the sources of the principal rivers ; but, with the exception of the Southern Alps, and some patches in the counties of Bathurst and Murray, this fundamental rock is visible in Australia, only where it appears to have cracked a thick overlying stratum of ferruginous sandstone. Thus, near the head of the river Cox, where the latter attains its greatest elevation ; and, from the character of the valley has evidently been violently disturbed; we find granite in the valley, near the bed of the stream.

… Such is the character of the country where the waters separate, or in the line of greatest elevation, which we are accustomed to term the Coast Range [the Great Dividing Range]. The general direction of this range is N. N. E. and accords perfectly with the hypothesis of Dr. Fitton, founded on the general parallelism observed in the range of the strata, even on the north-western coast, as noticed in his interesting little volume, the first ever devoted to Australian Geology. The parallelism so remarkable in the range of strata in that portion ; the general tendency of the coast-lines to a course from the west of south to the east of north, on the main land, and even in the islands west of the Gulf of Carpentaria ; and a general elevation of the strata towards the south-east as deduced from Flinders' remarks, are all facts which should be studied in connexion with the direction of the granite along this part of the eastern coast.

In *The Origin of Species*, Darwin noted that geologist, Reverend W.B. Clarke, saw the same "parallelism" extending to New Zealand and New Caledonia. Clarke had also noted that there was evidence of glaciation in the Australian Alps. These observations of Fitton, Mitchell and Clarke likely contributed to Wegener's Theory of Continental Drift.

… It may also be observed that the sandstone reposing on the rock eastward of the division or watershed, is slightly inclined towards the sea, whereas all the sandstone on the interior side, or westward of it, dips to the north-west.

Trap rocks[II] are displayed in a great variety of situations. They often occur connected with limestone in vallies ; sometimes constitute lofty ranges, as on the north or left bank of the Hunter ; and along the sea shore at the Illawarra ; they likewise cap the summit of isolated hills ; but no particular place can be assigned to them with reference to the position of any other rocks.

Trap forms a good soil on decomposition, as is shewn in the rich districts of the Illawarra, Cowpastures [Camden], Valley of the Hunter, Liverpool Plains, Wellington Valley, and Buree [Boree Creek west of Orange].

Vesicular lava and amygdaloid, are the chief ingredients of some of the best parts of Australia Felix [Victoria]. In that region volcanic phenomena are more apparent than in other parts of Eastern Australia, especially where the Grampians, consisting of a mass of sandstone 4000 feet thick, seem a portion of the great formation covering the districts of the north. The strata in these mountains are inclined to the north-west, as if in obedience to the upheavings of Murroa or Mount Napier, an extinct volcano, in the very line of their outcrop.

... We found in the interior, hills of sandstone only, but at this extremity [the south-western] of the great Coast range, granite is extensively exposed in ridges, between which, in one extensive district, are round heights of mammeloid form, consisting of pure lava : – and in another, tabular masses of trap reposing on granite, occupy one side of a valley. ...

Sandstone. The prevailing geological feature in all Eastern Australia, is the great abundance of a ferruginous sandstone in proportion to any other rocks. The sterility of the country where it occurs, has been frequently noticed in these volumes. It is found on the coast at Port Jackson, and it was the furthest rock seen by me, in the interior beyond the Darling.

A deposit, upwards of 1200 feet thick, forms the Blue Mountains west of Sydney – ranging thence with the intersection of no other

II Basic volcanic extrusive rocks such as basalt.

rock of importance, to the Hawkesbury ; and although declining towards the sea at the rate of only 100 feet per mile, or 1 in 52, or at an angle of about 1^0 with the horizon ; yet it is traversed by ravines, which increase in depth, in proportion as the sandstone attains a greater elevation, and present perpendicular crags and cliffs of a very remarkable character.

A region consisting of a sandstone deposit of so great a thickness and so slightly inclined, necessarily presents a monotonous aspect in all directions; and when it is compared with European countries composed of many formations, and presenting a great diversity of scenery, it proves how much geological structure influences pictorial and physical outlines.

Mitchell's observations of natural erosion put the piddling impacts of human activity in context:

The ravines which discharge their waters into the little river Cox, occupy an area of 1212 square miles … Supposing but two-thirds of the enclosed area of sandstone to have been excavated to the depth of 880 feet only, which I allow as the mean thickness of the stratum thus broken into, and considering the inclination of the Cox and other vallies, then, 134 *cubic miles* of stone [559 cubic kilometres] must have been removed from this basin of the Cox alone.

These days, one could be excused for thinking that erosion was a post-European invention. Indeed one of Australia's first naturist-scientists, Alec Costin, attributed natural erosion by nivation[III] as well as unnatural erosion after high-intensity post-European fires in the Australian Alps to grazing by domestic stock. Consequently, the reality of *The Man from Snowy River*[IV] became only a legend (Chapter 9).

Mitchell reported the discovery of fossilised bones of megafauna in the Wellington Caves:

III Loosening to erosion by water or wind of soil particles through alternate freezing and thawing of the soil surface under snowbanks.

IV A famous poem by A.B. 'Banjo' Paterson that later inspired a popular motion picture.

The pit had been first entered only a short time before I examined it by Mr. Rankin, to whose assistance in these researches, I am much indebted. He went down, by means of a rope, to one landing place, and then fixing a rope to what seemed a projecting portion of rock, he let himself down to another stage, where he discovered, on the fragment giving way, that the rope had been fastened to a very large bone, and thus these fossils were discovered.

He sent bones to Richard Owen, Professor at the Royal College of Surgeons, who identified giant kangaroos, wombats, possums and dasyurids (carnivorous marsupials), some of which were named after Mitchell. Owen was apparently torn between Creationism and Natural Selection but this didn't prevent him from trying to claim credit for the latter theory. Darwin wrote that:

> it appeared manifest to the Editor as well as myself, that Professor Owen claimed to have promulgated the theory of natural selection before I had done so; ... It is consolatory to me that others find Professor Owen's controversial writings as difficult to understand and to reconcile with each other, as I do.[31]

Owen alienated him to the extent that Darwin reputedly commented: "I used to be ashamed of hating him so much, but now I will carefully cherish my hatred and contempt to the last days of my life".

In addition to fossils, Major Mitchell and his native guides (outside their tribal country), discovered and described many living animals. For example "the two Tommies" captured the type specimen for Mitchell's hopping mouse, and "our native guides" caught the now extinct pig-footed bandicoot which Mitchell had previously encountered as a skull in Wellington Caves. Perhaps the best known animal associated with Mitchell's explorations is the admirable Major Mitchell cockatoo.

Long before ecologists became preoccupied with biodiversity, Mitchell made some interesting observations of the factors that govern its expression by comparing the environments near the Bogan and Darling Rivers:

> It has also its plains along the banks, some of them being very

extensive ; but the soil of these is not only much firmer, but is also clothed with grass and furnished with a finer variety of trees and bushes, than those of the Darling. Yet in the grasses, there is not such wonderful variety as I found in those on the banks of that river. Of twenty-six different kinds gathered by me there, I found only four on the Bogan, and not more than four other varieties, throughout the whole course. It appeared that where land was best and grass most abundant, the latter consisted of one or two kinds only, and, on the contrary, that where the surface was nearly bare, the greatest variety of grasses appeared, as if nature allowed more plants to struggle for existence where fewest were actually thriving.

He seamlessly melded biology, history, anthropology, sociology, paleoecology and geomorphology into a formidable ecology:

> Instead of the ferocious character latterly attributed to the natives of Van Diemen's Land [now Tasmania], we find, on the contrary, that Captain Cook describes them as having "little of that fierce or wild appearance common to people in their situation;" and a historian* draws a comparison, also in their favour, between them and the natives of Botany Bay, of whom three stood forward to oppose Cook, at his first landing. The ferocity subsequently displayed by natives of van Diemen's Land, cannot fairly be attributed to them, therefore, as characteristic of their race, at least until extirpation stared them in the face, and excited them to acts of desperate vengeance against all white intruders.
>
> *The History of New Holland, by the Right Hon. William Eden, 1787, p. 99.

He anticipated the debate whether Australian Aborigines have a single or multihybrid origin. This debate has recently been resurrected by Keith Windschuttle's interesting theory[32] of a conspiracy of silence around the pygmy Aborigines of North Queensland's rainforests. Mitchell noted that:

> Captain Cook … thus describes those [Tasmanian Aborigines] he saw at Adventure Bay in 1777: – "Their colour is a dull black, and

not quite so deep as that of the African negroes. It should seem also, that they sometimes heighten their black colour by smoking their bodies, as a mark was left behind on any clean substance, such as white paper, when they handled it." Captain Cook then proceeds to describe the hair as being wooly, but all other particulars of that description are identical with the peculiarities of Australian natives ; and Captain King stated, according to the editor of the northern voyage of Cook, that "Captain Cook was very unwilling to allow that the hair of the natives, we saw in the interior, and especially of the females, had a very frizzled appearance, and never grew long ; and I should rather consider the hair of the natives of Tasmania, as differing in degree only from the frizzled hair of those of Australia"...

The habits and customs of the Aboriginal inhabitants are remarkably similar, throughout the wide extent of Australia, and appear to have been equally characteristic of those of Van Diemen's Land: geological evidence also leads us to suppose, that this island has not always been separated from the main land by Bass's Straits. The resemblance of the natives of Van Diemen's Land to those of Northern Australia, seemed indeed so perfect, that the first discoverers considered them, "as well as the kangaroo, only stragglers from the more northern parts of the country;" and as they had no canoes fit to cross the sea, that New Holland, as it was then termed, "was nowhere divided into islands, as some had supposed". Their mode of life, as exhibited in the temporary huts made of boughs, bark, or grass, and of climbing trees, to procure the opossum, by cutting notches in the bark, alternately with each hand, as they ascend, prevails not only from shore to shore in Australia, but is so exactly similar in Van Diemen's Land, and at the same time so uncommon elsewhere, that Tasman, the first discoverer of that island, concluded, "that the natives either were of an extraordinary size, from the steps having been five feet asunder, or that they had some method which he could not conceive, of climbing trees by the help of such steps". It is strong presumptive evidence, therefore, of the connexion of the inhabitants of Van Diemen's Land with the race in Australia, that a method of climbing trees, now so well known as peculiar to the natives of Australia, should have been equally characteristic of those of Tasmania.

Mitchell would clearly favour the currently fashionable single origin theory rather than Birdsell's and Tindale's trihybrid theory.[33] The trihybrid theory had the North Queensland pygmies and the Tasmanian Aborigines as remnants of a negrito race that arrived earlier and then retreated from later invasions by Murrayian and Carpentarian races. Peter McAllister suggested that Windschuttle's motivation in resurrecting the Trihybrid Theory may be to put the European invasion in a better light as just one of a series. This would be an interesting twist on Flannery's line that puts Aboriginal and European invaders both in a bad light. McAllister argued that the Barrinean (North Queensland) pygmies were not much different to other Aborigines[34] as had Mitchell for the Tasmanians. But they **were** pygmies. This fact leaves 'hanging' the difficult questions of genetic drift over a short ten thousand years since rainforests reclaimed the Atherton Tablelands or alternatively, lingering phenotypic expression of malnutrition and disease since the pygmies came out of the rainforest and included more protein in their diet.[35]

Returning to Mitchell's broad-based ecology:

> The notches made in climbing trees, are cut by means of a small stone hatchet – and, as already observed, with each hand alternately. By long practice, a native can support himself with his toes, on very small notches, not only in climbing, but while he cuts other notches necessary for his further ascent with one hand, the other arm embracing the tree. The elasticity and lightness of the simple handle of the mogo or stone hatchet … are well adapted to the weight of the head–and assist the blow necessary to cut the thick bark with an edge of stone. As the natives live chiefly on the opossum, which they find in the hollow trunk or upper branches of tall trees, and, as they never ascend by old notches–but always cut new ones, such marks are very common in the woods ; and on my journies in the interior, I knew, by their being in a recent state, when I was approaching a tribe; or, when they were not quite recent, how long it had been since the natives had been in such

parts of the woods ; whether they had any iron hatchets, or used still those of stone only ; &c. …

The native dog, so common in Australia, is not found in Tasmania; while, on the other hand, two animals, the *dasyurus ursinus* [devil] and *thylacynus* [tiger], exist in Tasmania, but have not been found hitherto in Australia. Have these been extirpated in Australia by the dog, on his introduction subsequently to the opening of the straits? It may be observed that this is the more likely, as the above-mentioned species found in Van Diemen's Land only, consist of those two unable to climb and avoid such an enemy. … We find no remains of this genus among the fossils, and it seems therefore probable that the dog accompanied the native, wherever he came from.

With regard to dingoes, I would substitute "avoid such a competitor" for "avoid such an enemy" in the foregoing paragraph because I think the tiger and devil were supplanted on the mainland by a superior competitor.[36]

Mitchell compiled much geomorphological evidence of prehistoric changes in climate and sea levels including the separation of Tasmania from the mainland, drowned estuaries such as Port Jackson and Brisbane Water, former islands joined to the mainland by sand deposits such as Cape Solander and "Baranjuey", as well as wind formed crescent dunes on the northeastern margins of the salt lakes of the Murray-Darling Basin.

Consideration of linguistic similarities and differences amongst Aborigines led him tantalisingly close to describing another aspect of Australia's anthropological prehistory:

On comparing a vocabulary of the language spoken by the natives on the Darling, with other vocabularies obtained by various persons on different parts of the coast, I found a striking similarity in eight words, and it appears singular, that all these words should apply to different parts of the human body. I could discover no term in equally general use, for any other object, as common as parts of the body, such, for instance, as the sun, moon, water, earth, &c. By the … words used at different places to express the same meaning, it is obvious, that those to which I have alluded, are common to the natives both in the south-eastern and south-western portions of Australia ; while no such resemblance can be traced between these words and any in the language spoken by natives on the northern coast. Now, from this greater uniformity of language prevailing throughout the length of this large island, and the entire difference at much less distance latitudinally, it may perhaps be inferred, that the causes of change in the dialect of the aborigines, have been more active on the northern portion of Australia, than throughout the whole extent from east to west. The uniformity of dialect prevailing along the whole southern shore, seems a fact worthy of notice, as connected with any question respecting the origin of the language, and whether other people or dialects have been subsequently introduced from the northern or terrestrial portion of the globe.

The distribution of double-outrigger dugout canoes in Australasia also suggests close contact between northern Australian Aborigines and Indonesians or Melanesians[37] prior to European settlement. Mitchell was probably aware of Flinders' contact in northern Australia with a flotilla of Macassan trepangers under Chief Pobassoo who had visited Australia six or seven times over two decades. However Flinders' report of their apparent hostilities with the Aborigines was not suggestive of the history of economic, cultural and genetic transactions between these races dating from at least the early 1600s that are now known to have influenced northern Aboriginal phenotype, language and culture. Indeed Flinders considered and dismissed the possibility of such contact in the following terms:

> A question suggests itself here: could the natives of the west side of the Gulf of Carpentaria have learned the rite of circumcision from these Malay Mahometans? From the short period that the latter had frequented the coast, and the nature of the intercourse between the two people, it seems to me very little probable.[38]

In 1987 Stanford University linguist Merritt Ruhlen produced *A Guide to the World's Languages*. He classified Australian mainland Aboriginal

languages into two broad divisions similar to Mitchell's. Ruhlen
distinguished the Pama-Nyungan subgroup, which is found in the east,
south and west of the Australian mainland, and the non-Pama-Nyungan
languages of northern and northwestern Australia.[39] Keith Windschuttle
and Tim Gillin claimed this as evidence supporting the tryhybrid theory.[40]
But Mitchell's evidence connecting Tasmanian culture with the mainland
doesn't support the theory.

The following extracts from Mitchell's journal broadly describe the
landscape of southeastern Australia:

> … the soil in New South Wales is good only where trap, limestone,
> or granite rocks occur. Sandstone, however predominates so much
> as to cover about six-sevenths of the whole surface, comprised
> within the boundaries of nineteen counties. Wherever this is
> the surface rock, little besides barren sand is found in the place
> of soil. Deciduous vegetation scarcely exists there, no vegetable
> soil is formed, for the trees and shrubs being very inflammable,
> conflagrations take place so frequently and extensively in the
> woods during summer, as to leave very little vegetable matter to
> return to earth. On the highest mountains, and in places the most
> remote and desolate, I have always found on every dead trunk on
> the ground, and living tree of any magnitude also, the marks of
> fire ; and thus it appeared that these annual conflagrations extend
> to every place. …

> … This colony might thus extend northward to the Tropic of
> Capricorn, westward to the 145th degree of east longitude [close to
> the Paroo River] ; the southern portion having for boundaries, the
> Darling the Murray and the sea coast. Throughout the extensive
> territory thus bounded, one-third, probably, consists of desert
> interior plains ; one fourth of land available for pasturage or
> cultivation ; and the remainder, of rocky mountain, or impassable
> or unproductive country. Perhaps the greater portion of really
> good land, within the whole extent, will be found to the south-
> ward of the Murray, for there the country consists chiefly of trap,
> granite or limestone. …

> Every variety of feature may be seen in these southern parts, from
> the lofty alpine region on the east, to the low grassy plains, in which

it terminates on the west. The Murray, perhaps the largest river in all Australia, arises amongst those mountains, and receives in its course, various other rivers of considerable magnitude. These flow over extensive plains, in directions nearly parallel to the main stream, and thus irrigate and fertilise a large extent of rich country. Falling from mountains of a great height, the current of these rivers is perpetual, whereas in other parts of Australia, the rivers are too often dried up, and seldom, indeed, deserve any other name than chains of ponds.[41]

I had already been struck with the remarkable dissimilarity between the Murrumbidgee and all the interior rivers previously seen by me, especially the Darling. The constant fullness of its stream, its water-worn and lightly-timbered banks, and the firm and accessible nature of its immediate margin, unbroken by gullies, were all characters quite the reverse of those which I had seen elsewhere. Whatever reeds or polygonum might be outside, a certain space along the river, was almost everywhere clear, probably from its constant occupation by the natives. One artificial feature, not observed by me in other places, distinguishes the localities principally frequented by the natives, and consists in the lofty mounds of burnt clay, or ashes used by them in cooking.[42]

It was quite impossible to say on what part of the Murray, as laid down by Captain Sturt, we had arrived; and we were therefore obliged to feel our way, just as cautiously, as if we had been upon a river unexplored. The ground was indeed a tolerable guide, especially after we found, that this river also had bergs, which marked the line of separation between the desert plain or scrub, and the good grassy forest-land of which the river-margin consisted. As we proceeded, I found it best to keep along the bergs as much as possible, in order to avoid ana-branches* of the river.

*Having experienced on this journey the inconvenient want of terms relative to rivers, I determined to use such of those as recommended by Colonel Jackson in his able paper on the subject, in the Journal of the Royal Geographical Society for 1833, as I might find necessary. They are these:-

Tributary – Any stream adding to the main trunk.

Ana-branches – such as after separation unite.

Berg – bergs – Heights now at some distance, once the immediate banks of a river or lake.

Bank – that part washed by an existing stream.

Border – The vegetation at the water's edge, forest trees, or quays of granite &c.

Brink – The water's edge.

Margin – The space between the brinks and the bergs.

Where the bergs receded, forest land with the "goborro" or dwarf-box [*E. largiflorens*] intervened.[43]

Where there were no bergs, there were extensive floodplains carrying mosaics of grassland, woodlands of box, woodlands of "yarra" (river red gum, *E. camaldulensis*) and reedbeds. The next extracts refer to Mitchell's attempt to regain the riverbank at Gunbower Island which is now mostly red gum forest (Chapter 11):

> Mr. Stapylton found beyond the northern limits of the plain, amongst yarra trees, an ana-branch only, but containing quite clear and still water.
>
> The course of the creek, which I in the meantime traced, first led me to the north-east, where high trees seemed to mark its course … on returning to the main party, I found it desirable that the carts should cross. We next passed for three miles through a forest of goborro, and then a plain three miles in extent. Beyond the plain we approached a promising line of lofty yarra trees, but found it shaded only a hollow subject to innundations. Two miles and a half further we came to another similar line of trees, and we found within its shade an ana-branch full of clear water. A little in advance, a much deeper branch afforded a good spot for our camp, as I intended to cross it, by some means in the afternoon, and seek for the river. The plains we had crossed this day, were covered with excellent grass ; and in many places, detached groups of trees gave to the country a park-like appearance, very unlike anything on the banks of the Darling.
>
> After crossing the creek by means of a fallen tree, I found the ground beyond to be of the richest description, with excellent grass and lofty yarra trees growing upon it. I passed through two separate strips of high reeds, extending north-east and south-

west ; but I found they only enveloped lagoons of soft mud ; and seeing no appearance of the river at two miles from the camp, I returned.[44]

In other places, the river was bounded by sandhills and mallee (see front cover):

> The river taking a turn to the southward, we again entered the *dumosa* scrub, ... The soil consisted of barren sand ; there was no grass, but there were tufts of a prickly bush, which tortured the horses, and tore to rags the men's clothes about their ancles. I observed, that this bush and the *eucalyptus dumosa*, grew only where the sand seemed too barren and loose for the production of anything else ; so loose, indeed, was it, that, but for this dwarf tree, and prickly grass [*Triodia*], the sand must have drifted so as to overwhelm the vegetation of adjacent districts, as in other desert regions where sand predominates. Nature appears to have provided curiously against that evil here, by the abundant distribution of two plants, so singularly adapted to such a soil.[45]

Since European settlers cleared mallee in Victoria, dust storms have affected areas as far away as New Zealand.[46] Mitchell referred to what is now Victoria as "Australia Felix":

> Hills, of moderate elevation, occupy the central country between the Murray and the sea, being thinly or partially wooded, and covered with the richest pasturage, The lower country, both on the northern and southern skirts of these hills, is chiefly open ; slightly undulating towards the coast on the south, and is, in general, well watered.
>
> The grassy plains which extend northward from these thinly wooded hills, to the banks of the Murray, are chequered by the channels of many streams falling from them, and by the more permanent and extensive waters of deep lagoons. These are numerous on the face of the plains near the river, ...
>
> In the western portion, small rivers radiate from the Grampians, an elevated and isolated mass, ...
>
> Towards the sea coast on the south, and adjacent to the open

downs, between the Grampians and Port Phillip, there is a low
tract consisting of very rich black soil, ...

On parts of the low ridges of hills, near Cape Nelson and Portland
Bay, are forests of very large trees of stringy-bark, ironbark, and
other useful species of eucalyptus.[47]

The early explorers were looking for agricultural opportunities. They
had no reason to explore forests as we know them. However, Oxley, in
seeking a connection between the Northern Tablelands of modern New
South Wales and a coastal port, was forced to travel through tall, dense
forests on the escarpment above the Hastings River Valley. Heading east
from what is now the Walcha district, he was blocked by Apsley Gorge,
and headed in the general direction of the modern highway that bears
his name. He reached the escarpment forest at the edge of the Tableland
on 18th September 1818:

> During the night and this morning it has continued to blow a perfect
> equinoctial storm. ... Proceeding on our course to the east north-
> east, we did not advance above a mile and a half before a small
> stream running to the north-east through a very steep and narrow
> valley obliged us to alter our course more southerly, which we did,
> and soon entered a forest of stringybark and blue gum trees of
> immense size and great beauty. The soil on which they grew was a
> rich vegetable mould covered with fern trees and small shrubs. We
> found that this part of the country was intersected by deep valleys,
> the sides of which were clothed with stately trees, but of what
> kind we were ignorant [rainforest trees]: creepers and small timber
> trees, all of species not previously noticed by us, grew so extremely
> thick that we found it impossible to penetrate through them, We
> therefore continued along the edge of those valleys, our progress
> much impeded by the vast trunks of fallen trees in a state of decay,
> some of which were upwards of one hundred and fifty feet long,
> without a branch, as straight as an arrow, and from three to eight
> and ten feet in diameter. The forest through which we travelled
> appeared to be an elevated level or plain, and at three o'clock in the
> afternoon, after proceeding three or four miles to the westward, we
> cleared this truly primeval forest, and descended into a small valley
> of open ground, through which ran the stream we had crossed

in the morning. Indeed we were not more than two miles south of the place we had quitted. Our hope of proceeding without much interruption was thus disappointed: the gloominess of the weather, and the constant showers that fell, so impeded our view and distorted its objects, that what appeared plain and practicable at a distance of two or three miles, when approached was found impassable. I think it probable, however, that our most serious obstructions will be the thickness of the timber, rotten trees, and creeping plants; the soil is so rich and free from rocks, that I do not think the steepness of the descents will greatly endanger us.[48]

The "truly primeval forest" was a sharp contrast to the "forest land" that Oxley found upon descending into the Hastings Valley:

> The country we passed through is what is generally known in New South Wales as open forest land, with occasionally small flats on the river: steep hills sometimes ended on the river, and north and south of us were detached ranges of a similar description. The whole face of the country was abundantly covered with good grass, which having been burnt some time, now bore the appearance of young wheat.[49]

Mitchell concluded his description of Victoria as follows:

> The high mountains in the east, have not yet been explored, but their very aspect is refreshing, in a country, where the summer heat is often very oppressive. The land is, in short, open and available in its present state, for all the purposes of civilised man. We traversed it in two directions with heavy carts, meeting no other obstruction than the softness of the rich soil ; and, in returning, over flowery plains and green hills, fanned by the breezes of early spring, I named this region Australia Felix, the better to distinguish it from the parched deserts of the interior country, where we had wandered so unprofitably, and so long.

The gap in knowledge about the high mountains was quickly filled by the explorer Strzelecki, the botanist Mueller, Deputy Surveyor-General Townsend and geologist-clergyman W.B. Clarke.[50]

> The Sub-Alpine Vegetation buried in frost during the winter

months, stretches from 4000 to 5000 feet in elevation above the sea [1200-1500 metres]. The trees and shrubs which it produces are of diminutive growth, and with their cessation, or with their reduction to a very stunted and prostrate condition, begins the truly Alpine Flora, which commences at an average elevation of 5000 feet, although, according to the position of the mountains and their isolation and approximation to other ranges, Alpine plants may be seen descending in some localities fully a thousand feet lower than in others. On the very crest of Mount Hotham, of Mount Latrobe, and on Mount Kosciusco [sic], and several others of the Muniong Mountains, there exist, even, no shrubs of depressed growth, the cessation of which takes places at about 6000 feet of elevation [1800 metres].[51]

Immediately after descending from this region of eternal snow the main range forms a very deep gap, falling about (?) two thousand feet to the source of the Crack-em-back River; it is there covered with stunted gum trees– gnarled and twisted by the wind in such a manner as to resemble mangroves, than which they seldom attain a greater height. … About the source of the Toonginbooka the character of the timber alters. There are some small portions of forest in which grows the large Eucalyptus tree known as "Black," but surrounded by "tea tree," with occasional patches of bitter leaf [*Daviesia*] or willow scrub, which presents an obstacle almost impassable to man. The summits of the peaks were generally more densely clad than any other part, making it impracticable to get a view whence I could make out the general run of the range …

The blacks had visited the Snowy Mountains, a short time previously to us, for the purpose of getting "Bogongs," a species of moth, about an inch long, of which they are particularly fond; to obtain them they light large fires, and the consequence was, the country throughout the whole survey was burnt, leaving my bullocks destitute of food. During the time I was on the range the lower parts of the country were burning, and I was prevented, in almost every instance, from getting angles on any distant points, by the dense masses of smoke obscuring the horizon in all directions. It is also worthy of remark that not a single head of cattle, kangaroo, or emu, was seen during the whole journey along the range, and only a very few old hacks.[52]

As Mr. Townsend mentions the scarcity of kangaroos in the Alps, a fact I have also noted above, I may also mention that he told me that he had named one part of the Muniong Range "Snake Hill," from the abundance of that reptile. I also found brown snakes, of large size, very common in Alum Creek, in October, 1851; I was also often warned against entering the long grass on the banks of Creeks in the Alpine country, where they are said to abound. Scarcely any animals, but snakes and rats, are found in the interior desert of Australia; and rats are very plentiful at the head of the Eucumbene.[53]

After his final expedition northwards *into the interior of Tropical Australia*,[54] Mitchell succinctly emphasised the role of human fire in Australian ecology:

Fire, grass, kangaroos, and human inhabitants, seem all dependent on each other for existence in Australia ; for any one of these being wanting, the others could no longer continue. Fire is necessary to burn the grass, and form those open forests, in which we find the large forest-kangaroo ; the native applies that fire to the grass at certain seasons, in order that a young green crop may subsequently spring up, and so attract and enable him to kill or take kangaroo with nets. In summer, the burning of long grass also discloses vermin, birds' nests, &c., on which the females and children, who chiefly burn the grass, feed. But for this simple process, the Australian woods had probably contained as thick a jungle as those of New Zealand or America, instead of the open forests in which the white men now find grass for their cattle.

During this expedition to the tropics Mitchell was obstructed on seven occasions by fallen timber in acacia scrubs. Once he noted that:

Our carts had been so much jolted about and shaken, in crossing the dead timber yesterday, that I resolved to keep along the river bank this day . . . there was less fallen timber, as the natives had been accustomed there to make their fires, and roast the mussles of the river, and other food…. The fallen timber of the brigalow

[*Acacia harpophylla*] decays very slowly, and is not liable to be burnt, like most other dead wood in open forests, because no grass grows amongst the brigalow, as in open forests.

Mitchell's surveys, like Townsend's, were sometimes impeded by scrubs on rocky summits that weren't burnt by Aborigines because there was no grass to carry fire:

> In hopes of obtaining an elevated view over the country to the westward, I endeavoured to ascend the northern summit of Mount Abundance, but although the surface to near the top was tolerably smooth, and the bush open, I was met there by rugged rocks, and a scrub of thorny bushes so formidable as to tear leathern overalls, and even my nose. After various attempts, I found I was working round a rocky hollow, somewhat resembling a crater, although the rock did not appear to be volcanic. The trees and bushes there were different from others in the immediate vicinity, and, to me, seemed chiefly new. It is, indeed, rather a curious circumstance, but by no means uncommon, that the vegetation on such isolated summits in Australia, is peculiar and different from that of the country around them. Trees of a very droll form chiefly drew my attention here [bottle trees – *Brachychiton rupestris*]. The trunk bulged out in the middle like a barrel, to nearly twice the diameter at the ground, or of that at the first springing of the branches above. These were small in proportion to their great girth, and the whole tree looked very odd. These trees were all so alike in general form that I was convinced this was their character, and not a lusus naturae [freak of nature]. (A still more remarkable specimen of this tree was found by Mr. Kennedy [Assistant Surveyor] in the apex of a basaltic peak, in the kind of gap of the range through which we passed on the 15th of May, and of which he made the accompanying drawing.) (Fig. 3, p. 71.)
>
> These trees grew here only in that almost inaccessible, crater-like hollow, which had impeded me in my attempt to reach the summit. Leaving the horses, however, I scrambled through the briars and up the rocks to the summit, but found it, after all this trouble, too thickly covered with scrub to afford me the desired view to the westward, even after I had ascended a tree on the edge of the

Fig. 3 Mr Kennedy's drawing of a bottle tree
(Courtesy Archive Digital Books Australasia)
broad and level plateau, so thickly covered with bushes.

When Curr took up country at Tongala on the Goulburn River he saw the wheel ruts left in the bare soils of the plains six years earlier by Mitchell's southern expedition. Curr described the fundamental role of Aboriginal fire in Australia,[55] effectively dispatching *The Future Eaters* hypothesis a century before Flannery constructed it:

> Mere hunters, who absolutely cultivated nothing—the spear, the net, the tomahawk—could have produced no appreciable effect on the natural products of a large continent. Nor did they ; but there was another instrument in the hands of these savages which must be credited with results which it would be difficult to over-estimate. I refer to the *fire-stick*; for the blackfellow was constantly setting fire to the grass and trees, both accidentally, and systematically for hunting purposes. Living principally on wild roots and animals, he tilled his land and cultivated his pastures with fire; and we shall not,

perhaps, be far from the truth if we conclude that almost every part of New Holland was swept over by a fierce fire, on average, once in every five years. That such constant and extensive conflagrations could have occurred without something more than temporary consequences seems impossible, and I am disposed to attribute to them many important features of nature here; for instance, the baked, calcined, indurated condition of the ground so common to many parts of the continent, the remarkable absence of mould which should have resulted from the accumulation of decayed vegetation, the comparative unproductiveness of our soils, the character of our vegetation and its scantiness, the retention within bounds of insect life (notably of the locust, grasshopper, caterpillar, ant and moth), a most important function, and the comparative scarcity of insectivorous birds and birds of prey. They must also have had an influence on the thermometrical range, and probably affected the rainfall and atmospheric and electrical conditions.

When these circumstances are weighed, it may perhaps be doubted whether any section of the human race has exercised a greater influence on the physical condition of any large portion of the globe than the wandering savages of Australia.

Curr deduced that the size of the Aboriginal economy on the Murray River, and its environmental impact, had been lessened in advance of the wave of European settlement by a smallpox epidemic. Of the ongoing decline of the Bangerang tribe at Tongala (the Bangerang name for the Central Murray) he wrote:

> Speaking of this subject brings to mind a circumstance (and there were possibly others now forgotten) which led me, in those days, to suspect that the Bangerang and neighbouring tribes had greatly fallen off in numbers some time prior to the settlement of the whites in Port Phillip [Melbourne].
>
> At Colbinabbin and Coragorag, in the country of the Pinpandoor, one met with strong evidence of such having been the case. I refer to the ovens found in their country, which were out of all proportion numerous when compared to the wants of the tribe as it then existed. The prominent feature in connection with these ovens,

many of which, at the time of which I write, had long been disused, was, not that their capacity for cooking as regards size was in excess of what the tribe would require (as this would at last inevitably result from constant addition of material), but that very commonly two or three large ovens existed in close proximity, which, in view of the customs of the tribe, I can only account for on the supposition of much larger numbers of Blacks than we ever saw, having in former days habitually assembled at these places – numbers whose food it would be inconvenient to cook in one oven.

Though the dimensions of each oven were not so important to his deductions, they were certainly impressive. Curr measured four ovens around the Paiangil head station. The largest diameter was 23 metres and the highest was 1.5 metres.

> The conclusions to which the facts connected with these ovens seem to lead are, first, that in 1841 the Blacks in the neighbourhood in which I lived, were not so numerous as they had been ; and second, to judge by the trees growing out of the ovens, that the reduction in their numbers had taken place some fifty years or so antecedent to that date. As to the cause of such reduction, there are strong reasons for assigning it to smallpox, of which disease, in its most virulent form, many individuals amongst the Pinpandoor, Ngooraialum, Bangerang and other neighbouring tribes bore the marks. Indeed, I and several others, two of whom are still alive, saw at Tongala in 1843, or thereabouts, a Bangerang child absolutely suffering from this complaint.

The chronology that Curr deduced from the size of the trees would tally with a rapid spread of disease from Port Jackson after 1788. However Flood argued that smallpox could not have arrived with the first fleet because there were no active cases and it has an incubation period of a fortnight compared to the fleet's journey of months from the previous potential source. She suggested that it was introduced to Australia by Macassan trepangers during the 1780s when there were epidemics in the Sunda Islands.[56]

Smallpox never reached Tasmania, but an epidemic disease, probably influenza contracted from visiting sailors or sealers, decimated the Aboriginal population before European settlement. In 1831, Protector of Aborigines, George Augustus Robinson saw the consequences in northeastern Tasmania: "many of those districts which had been formerly peopled by the Aborigines are now unoccupied; the once resident tribes being utterly extinct, a fact which was evinced by the dense overgrown underwood".[57]

Curr described mosaic burning by Aborigines in mallee near Lake Boga and in reedbeds at Moira. Flinders,[58] on the other hand, described the vegetation and fire regimes of wilderness islands off our southern shores:

> 21 February [1802] ~ Early in the morning, I went on shore ... Signs of extinguished fire existed everywhere; but they bespoke a conflagration of the woods, of remote date, rather than the habitual presence of men, and might have arisen from lightning ...Upon the whole I satisfied myself of the insularity of this land; and gave to it, shortly after, the name of Thistle's Island, from the Master who accompanied me. ... a white eagle, with fierce aspect and outspread wing, was seen bounding towards us; but stopping short, at twenty yards off, he flew up into a tree. Another bird of the same kind discovered himself by making a motion to pounce down upon us as we passed underneath; and it seemed evident that they took us for kangaroos, having probably never before seen an upright animal in the island, of any other species. ...
>
> In expressing an opinion that these people have no means of passing the water, it must be understood to be a deduction from our having met with no canoe, or the remains of any about the port [Port Lincoln]; nor with any tree in the woods from which a sufficient size of bark had been taken to make one. Upon Boston Island, however, there were abundant marks of fire; but they had the appearance, as at Thistle's Island, of having been caused by some conflagration of the woods several years before ...
>
> Neither smokes, nor other marks of inhabitants had as yet been perceived upon the southern land, although we had passed along seventy miles of its coast. ... next morning, however, on going

towards the shore, a number of dark-brown kangaroos were seen feeding upon a grass plat by the side of the wood … I killed ten, and the rest of the party made up the number to thirty-one …

After this butchery, for the poor animals suffered themselves to be shot in the eye with small shot, and in some cases to be knocked on the head with sticks, I scrambled with difficulty through the brush wood, and over fallen trees, to reach the higher land with the surveying instruments; but the thickness and height of the wood prevented anything else from being distinguished. …

… I named this southern land Kangaroo Island. …

The soil of that part of Kangaroo Island examined by us, was judged to be much superior to any before seen, either upon the south coast of the continent, or upon the islands near it … Some sand is mixed with the vegetable earth, but not in any great proportion;

He deduced that high-intensity fires on these wilderness islands had been ignited by lightning, and he contrasted their fire regimes and soils against those on the mainland (Chapter 8). Mitchell[59] had remarked upon the absence of vegetable earth on the Sydney sandstones where Aborigines constantly burnt vegetation.

I hope that the sample of historical tidbits in this chapter might be sufficient to whet the appetite, so that readers are tempted to savour the smorgasbord of historical ecology that is now readily available electronically and in rebirthed books. It is very instructive to focus on a region and build up a detailed qualitative picture of the pre-European landscape, then compare it against the picture based on 'quantitative data' from 'reference stands' that is favoured by many theoretical ecologists. I have done this in two papers[60] about woodlands in south-eastern Australia. Anyone interested in practical conservation or restoration of biodiversity, should study the relevant ecological history before deciding on the best approach. It will often be diametrically opposed to what is recommended in modern ecological literature or enshrined in environmental legislation.

Endnotes

1 Horton 1982, Benson and Redpath 1997.

2 Tench 1793.

3 White 1790.

4 Mitchell 1839.

5 USA Today 2013.

6 Meiklejohn *et al.* 2013.

7 Price 1798.

8 Campbell 1926, p. 40.

9 Geographical Names Board 1970.

10 Blaxland 1823.

11 Lawson 1813.

12 Wentworth 1813.

13 NSW Government 1814.

14 Evans 1814.

15 Flinders 1814.

16 Evans 1815.

17 Bathurst 1815.

18 Oxley 1820.

19 Sturt 1833.

20 Mitchell 1839, 1848.

21 Mitchell 1839.

22 Darwin 1845.

23 Curr 1883.

24 Noble 1997.

25 Calaby 1983, p. 176.

26 Darwin 1860.

27 Flood 2006.

28 op. cit.

29 op. cit.

30 Mitchell 1839, Vol. II, pp. 355-8.

31 Darwin 1860.

32 Windschuttle and Gillin 2002.

33 Tindale and Birdsell 1941.

34 McAllister 2010.

35 Ibid.

36 Newsome 1983.

37 Hornell 1943.

38 op. cit.

39 Ruhlen 1987.

40 op. cit.

41 Mitchell 1839, Vol. II, pp. 328-31.

42 Ibid., pp. 80-1.

43 Ibid., pp. 90-1.

44 Ibid., p. 150.

45 Ibid., pp. 106-7.

46 *The Examiner* 1944.

47 op. cit., pp. 331-2.

48 op. cit.

49 op. cit.

50 Clarke 1860.

51 Mueller 1855.

52 Townsend 1846.

53 Clarke 1860.

54 Mitchell 1848.

55 op. cit.

56 op. cit.

57 Flood 2006.

58 op. cit., pp. 81-94.

59 Mitchell 1839.

60 Jurskis 2009, 2011a.

3a
HOW THINGS WERE AGAIN

When I see a bird that walks like a duck and swims like a duck and quacks like a duck, I call that bird a duck.

James Whitcomb Riley

Many accounts by early Australian explorers show how geological processes such as the Kosciuszko Uplift, volcanism, erosion and sedimentation, as well as climate variability, shaped our topography and soils. The pre-European landscape resulted from the interaction of topography and soils with ongoing climate change, biota and the firestick. These interactions were many and varied but there was mostly no mystery.

Flat to gently undulating country inland of the Great Dividing Range, and rainshadow areas in broad coastal river valleys supported savanna or grassy woodland. These were the areas Mitchell classed as "land available for pasturage and cultivation". He estimated that this type of country occupied 25% of the area bounded by the Tropic of Capricorn, the Paroo River, the Darling, the Murray and the sea.[1] Eucalypts almost exclusively dominated subhumid areas. In drier areas they occurred in mixtures with cypress, casuarinas, various acacias and other small trees. Sandhills in the semi-arid zone were occupied by cypress (*Callitris*). Aboriginal burning maintained open woodlands with scattered mature trees and shrubs, various herbs and grasses interspersed with bare ground, scattered lignotubers and few saplings.

Cypress is extremely tolerant of drought. It has a compact crown and relatively low turnover of foliage. It seeds prolifically and, in wet seasons, seedlings escape competition from grasses as their roots penetrate deeply into the soil profile.[2] Cypress is well adapted to sandhills where seedlings rapidly access water at great depth. However rapid root growth occurs at the expense of shoot growth. Eucalypts have spreading crowns and relatively rapid turnover of foliage. Lignotuberous eucalypts are less than-prolific seeders, and seedlings initially direct their energy towards production of a lignotuber that is resistant to fire and drought, rather than producing deeply penetrating roots. They have an advantage on heavier plains soils where water retention at the surface is stronger, and fires were more frequent.

Seedling regeneration of belah (*Casuarina cristata*) is reported to be very rare, though it is easily propagated from seed in nurseries. Belah mostly regenerates from root suckers.[3] Whatever the physiological explanation, there are obvious trade-offs, in drought adapted species, between prolific seedling regeneration and vegetative reproduction or persistence. River red gum in southern Australia is not lignotuberous and is similar to cypress in its recruitment pattern. It seeds prolifically and seedlings rapidly extend their roots down into sandy aquifers connected to watercourses.[4]

Historically, Aboriginal burning destroyed nearly all woody seedlings, favouring herbs and grasses. After floods or wet seasons there was prolific growth of grass or reeds, prolific germination of cypress, river red gum and bulloak (*Allocasuarina luehmannii*), sporadic germination of lignotuberous eucalypts and very limited germination of belah. When grass cured, Aborigines burnt it, destroying most cypress, red gum and bulloak seedlings, killing lignotuberous eucalypt seedlings back to ground level and regenerating herbs and grasses. They were able to burn mixed woodlands more frequently than cypress stands on sandhills, because growth of herbage is more reliably abundant on the plains and litterfall from eucalypt trees is much greater.

Occasionally, woody seedlings grew to shrubs or saplings where they initially escaped fire and competition from grasses, in refuges such as bare ground or ashbeds from burnt fallen timber. Lignotuberous eucalypt seedlings are intolerant of competition from established trees. They grew to saplings when they were simultaneously released from overstorey competition and from the 'fire trap', as for example, when a hollow-butted old tree burnt down and its green remains, or an ashbed where it partly burnt away, protected the seedling from fire for at least one season.

On floodplains, there were lines, strips and small patches of river red gum. These grew on high banks along rivers and runners (small ephemeral ana-branches), around lagoons and in bends of rivers, where underlying aquifers were recharged by high flows. Low floodplains behind banks or between runners, were flooded every year and carried reedbeds and swamps. Open woodlands occupied higher flood plains where underlying aquifers were not so well connected to river flows and floods were not as frequent. Low floodplains on the Lachlan and the Darling didn't receive the benefits of alpine precipitation and were flooded irregularly. They mostly grew "polygonum" (*Muehlenbeckia*) and box (*Eucalyptus*).

Some extensive scrubs occurred in the country between the headwaters of the Darling River and the headwaters of the Burdekin in Central Queensland. These were mixed in together with the land that Mitchell classed as "available for pasturage or cultivation". However many brigalow (*Acacia harpophylla*) scrubs were littered with fallen timber, and, together with some other scrubs, were sometimes so densely treed or shrubbed as to impede the progress of Mitchell's expedition.[5] There was no grass to carry fire through these stands, but the soils are capable of growing grass. This raises a question whether the scrubs are a remnant of the pre-Aboriginal vegetation that human fire could not penetrate or whether they represent an unusual case where the advent of human fire created something other than a savanna.

In *Burning Bush* Stephen Pyne favoured the former explanation. He focused on brigalow, noting that its leaf litter doesn't make good fuel and that:

> Leaf fall, moreover, is keyed to intermittent rains rather than to seasons, complicating the timing of kindling to spark routine fire. … For practical purposes, combustible litter does not exist … There is little to compel the brigalow belt to accept fire.[6]

But it's not just brigalow that forms impenetrable scrubs in the brigalow belt. Mitchell's expedition was hindered by rosewood (*Heterodendrum oleifolium*), cypress, mulga (*Acacia aneura*), belah and mixed scrubs in this region. Scrubs occurred on all types of soil including sand, claypans and hard ridges. They were intimately mixed with "very open forest" and with "good grassy land".

Most importantly the extensive scrubs were ephemeral. Mitchell found many fine grassy areas with standing or fallen dead brigalow, or merely stumps remaining. For example:

> I continued to seek the river across extensive downs, in many parts of which dead brigalow stumps remained. … An uncommon drought had not only dried up the waters of this river, but killed much of the brigalow scrub so effectually, that the dead trunks alone remained on vast tracts, thus becoming open downs.[7]

I think that the answer to the question of their origin lies in the intermittent rains and complicated timing of fire. The scrubs most probably established after sustained wet periods stimulated mass germination of woody seedlings. Grasses may not have sprouted and cured in time to for Aboriginal fires to destroy the woody seedlings. Or perhaps Aborigines weren't there at the right time to burn.

Smallpox epidemics originating from the north coast may have decimated the tribes in the brigalow belt at the same time as wet conditions promoted woody seedling establishment. Brigalow is very drought-tolerant. Mitchell saw that mass mortalities had occurred in dense scrubs of brigalow, but not in "forest land". He didn't see shifting boundaries between trees and grass according to sustained drought or waterlogging as he did in other regions. Drought deaths are common in the post-European forests that replaced drought-resistant open woodlands when Aboriginal burning was disrupted.

Extensive dense scrubs may be an unnatural phenomenon arising from disruption of Aboriginal management by disease. Certainly smallpox would have been more virulent near the tropics and would have become less deadly as it moved south through the Aboriginal population, spread by 'carriers' who survived infection.[8] Even so, Curr described trees and scrubs that grew up after local tribes were decimated by a smallpox epidemic on the Murray.[9] In northeastern Tasmania, Robinson saw "dense overgrown underwood" that had invaded woodlands after Aborigines were decimated by influenza.[10] Gammage suggested that "At the risk of their souls, people cared for every inch of their country, even after smallpox halved their numbers and management capacity".[11] Robinson's and Curr's observations show that this was not the case.

Brigalow and belah sucker freely from the roots – possibly an adaptation to shearing of the roots by swelling and cracking of heavy clay soils. Thus they have an innate tendency to form scrubs on claypans[12] where Mitchell often encountered them. In many other situations, prolific growth of dense suckers and seedlings after early rounds of Aboriginal firing may have allowed scattered trees to form clumps that were impervious to subsequent mild fires. The clumps would suppress other trees and grasses as do post-European cypress scrubs.[13] Clumps may have coalesced into patches of scrub. But it is unlikely that extensive scrubs would have developed unless Aboriginal management was substantially disrupted for a considerable time.

Thirty-three percent of "the colony" as defined by Mitchell was "desert interior plains" or rangelands in modern terminology. Here drought-tolerant eucalypts usually occupy the heavier soils and drainage depressions while cypress or trees of smaller stature such as rosewood, belah or mulga may dominate lighter, stonier or drier soils. Mallee often dominates dunes in the arid zone together with porcupine grasses (*Triodia*), whilst spear grass (*Stipa*) occupies the swales where moisture occasionally accumulates and severe frosts sometimes occur. Aborigines were able to burn the cured grasses in the swales more often than the dunes where fire would only run after sufficient porcupine grass and mallee litter had accumulated during an unusually wet season or several average seasons.

Forty-two percent of 'the colony' was "rocky mountain, or impassable or unproductive country". Much of this is very rocky or very dry or very wet country where fire cannot run freely because growth of herbage is usually too sparse or too wet. Depending on the particular soil conditions, very difficult sites for trees and/or grasses, including very dry, very exposed, solid rock, waterlogged, frost hollow and alpine sites, support grasslands, sedgelands, wetlands, shrublands, heaths or scrubs. Some deserts support spinifex (*Triodia*) grasslands.

Various mallees or acacias or other, often dense and prickly bushes dominate ridges and hills that are too stony to grow grass and carry mild fire. When sufficient litter had accumulated over several years, and there was enough heat and wind, Aborigines were able to burn these scrubs with moderate or high-intensity fires as they did in the Blue Mountains in January 1814.[14] They could do so safely because the fires died down when they ran into open country with green grass or bare ground or ashes from recent grass fires. Further south at higher altitudes, Aborigines visited the Alps annually to gather bogong moths. They lit fires in grasslands and woodlands. There were no kangaroos or wallabies to hunt, so there was no need to burn in a fine mosaic pattern. Surveyor Townsend noted that their fires were very extensive, and he called the tree we now know as alpine ash (*E. delegatensis*), black gum, obviously for the colour of its rough bark when regularly singed by fire. But these mild fires didn't penetrate alpine scrubs of "tea tree" and "bitter leaf".[15]

Eucalypt forests grow on less stony mountain country where trees have an advantage over grasses because their roots can access water down in the regolith.[V] Enough fuel accumulates as herbage and litter to carry moderately frequent fire but shading by topography and canopy are such that fuels are often too moist to carry a mild fire. Aborigines burnt forests less frequently than savannas because there was less opportunity. Forests were open and grassy with more or less scattered shrubs except in very moist, sheltered positions which were infrequently dry enough to carry fires. Taller eucalypts with understoreys of shrubs or treeferns occupied these positions. Here, growth of shrubs was controlled by established canopy and recurrent fire as in woodlands. Infrequent fires of moderate intensity killed the shrubs but not the canopy trees. Some shrubs resprouted and some grew back from seed.

V Loose unconsolidated rock fragments and soil laying on bedrock.

Rainforests took refuge on relatively fertile, well structured and well watered soils where their tall, dense canopies excluded sunshine and wind so that combustible fuels did not accumulate. Oxley described such a situation on a high-rainfall plateau of deep ferrosol where huge rotten logs impeded his progress. Aboriginal fire could not penetrate, so Aboriginal economy was excluded from this "truly primeval forest".[16] The exception was in North Queensland where a large area of diverse rainforest allowed a few tribes to develop a specialised economy based on fruits and nuts of rainforest trees. Without fire, grass and kangaroos across the landscape, these tribes apparently lacked protein in their diet. Their way of life and diet were so different that they somehow became pygmies, phenotypically distinct and possibly isolated genetically from other Aborigines.[17]

Slightly drier or more exposed positions in rainforest areas are occupied by tall, wet eucalypt forests with a subcanopy of rainforest trees. These forests are not flammable except under extreme weather conditions. Naturally infrequent high-intensity fires temporarily destroy the canopy and subcanopy, allowing the highly intolerant (of shade), fast-growing, light-demanding eucalypts to regenerate from seed and quickly establish a new canopy under which the shade-tolerant subcanopy species develop. Rainforests burnt perhaps once in a thousand years, and wet eucalypt forests once every few centuries,[18] when extreme climatic and weather conditions coincided with a lightning ignition, or delivered embers to the elevated fuels in the wet forest. As may have happened with pygmies, wet sclerophyll eucalypts growing amongst extensive rainforests became genetically isolated.[19]

Fire was critical to Aboriginal economies, and Aborigines enhanced the vegetation mosaic by using differences in flammability according to season and site. They burnt smaller patches where resources were rich or diverse (e.g., in the Bunya Mountains or the Western Deserts[20]) and larger patches where resources were fewer or more homogenous (e.g., in the Alps or in jarrah (*E. marginata*) forests[21]). Whatever the patch size, they created a shifting mosaic of green, brown and black – new growth, bare ground and freshly burnt vegetation.

Aboriginal burning favoured animals and plants that occupied open grassy habitats, and confined other plants and animals to physically sheltered refuges from fire, or patches that had been burnt a few years before. Some animals nested, rested or hid from predators in such refuges, and foraged in the open. Plants and animals that relied on decaying litter and timber in moist shady sites for food and shelter, were confined to rainforests, tall wet eucalypt forests, closed scrubs and dark gullies.

Plants and animals were attuned to a highly variable climate and cycles of boom and bust. Plants mostly survived extremes of drought or waterlogging, though vegetation boundaries might shift according to medium scale climatic cycles as Mitchell[22] described of woodland/wetland/grassland boundaries. Plants and animals produced relatively large numbers of offspring in unusually favourable seasons but, apart from very mobile animals or lignotuberous plants, most did not survive to maturity because they were burnt after flushes of green vegetation cured, or they perished in subsequent dry seasons, or they failed to develop before ephemeral waterbodies disappeared. Prolific breeding of pelicans when Lake Eyre occasionally fills and massive mortality of chicks as the lake dries provides an example of natural boom and bust.

A number of short, medium and long term cycles govern earth's orientation and proximity to the sun, as well as fluxes of energy from the sun. Australia's location between the tropics and the southern ocean, and between the Indian and Pacific Oceans as well as its size – the largest island and smallest continent – together with these various cycles readily explain the extreme climatic variability noted by the early explorers and recently rediscovered by global warming enthusiasts.

Endnotes

1 Mitchell 1839.
2 Lacey 1973.
3 Cunningham *et al.* 1981.
4 Jacobs 1955.
5 Mitchell 1848.

6 Pyne 1991.

7 Mitchell 1848.

8 Flood 2006.

9 Curr 1883.

10 Flood 2006.

11 Gammage 2011.

12 Cunningham *et al.* 1981.

13 Jurskis 2009.

14 Evans 1814.

15 Townsend 1846.

16 Oxley 1820.

17 McAllister 2010.

18 Turner 1984.

19 Jones *et al.* 2006.

20 Burrows *et al.* 2006, Gammage 2011.

21 Townsend 1846, Abbott 2003.

22 Mitchell 1839.

4

HOW THINGS CHANGED

A child, however, who had no important job and could only see
things as his eyes showed them to him, went up to the carriage.
"The Emperor is naked" he said.

Hans Christian Andersen

Despite much obfuscation by naturists, there's no mystery as to how
and why things changed after European settlement. A.W. Howitt
F.G.S., F.L.S. used examples from Gippsland to explain the fundamental
cause when he read a paper before the Royal Society of Victoria in
1890:[1]

> The influence of settlement on the Eucalyptus forests has not been
> confined to the settlements upon lands devoted now to agriculture
> or pasturage, or by the earlier occupation by a mining population.
> It dates from the very day [in the 1830s] when the first hardy
> pioneers drove their flocks and herds down the mountains from
> New South Wales into the rich pastures of Gippsland.
> Before this time the graminivorous marsupials had been so
> few in comparative number, that they could not materially affect
> the annual crop of grass which covered the country, and which
> was more or less burnt off by the aborigines, either accidentally
> or intentionally, when travelling, or for the purpose of hunting
> game.
> These annual bush fires tended to keep the forests open, and to
> keep the open country from being overgrown, for they not only
> consumed much of the standing or fallen timber, but in a great
> measure destroyed the seedlings which had sprung up since former
> conflagrations.
> The influence of these bush fires acted, however, in another

direction, namely, as a check upon insect life, destroying, amongst others, those insects which prey upon the eucalypts.

Granted these premises, it is easy to conclude that any force which would lessen the force of the annual bush fires, would very materially alter the balance of nature, and thus produce new and unexpected results. ...

The older trees of this second growth do not, I suspect, date further back than the memorable "Black Thursday," [sic] when tremendous fires raged over this tract of country. It may also be inferred, from the constant discoveries during the process of clearing of blackfellows' stone tomahawks, that much of this country, now covered by a dense scrub of gum saplings, Pomaderris apetala, Aster argophylla, and other arborescent shrubs, that the country was at that time mainly an open forest.

Disruption of Aboriginal burning certainly altered the balance of nature with unexpected consequences, but Howitt didn't fully understand the underlying processes. Aboriginal burning killed woody seedlings, burnt cured litter and herbage, released some nutrients to the atmosphere and cycled some nutrients back to established trees or herbs and grasses. Mitchell described how Aborigines used green pick to attract kangaroos, and Curr described Aboriginal burning to promote manna in the mallee around Lake Boga.[2]

Aboriginal burning created flushes of new growth and of animals that eat it, but prevented sustained irruptions of plants and animals. It controlled woody seedlings by killing them, but it controlled arbivores indirectly – not by killing them as has been suggested in respect of insects, koalas and mistletoes.[3] Curr recognised that disruption of Aboriginal burning had affected nutrient cycling and soil conditions[4] but despite his comments about Aborigines farming manna he did not recognise the link between burning and the nutritional value of trees to their arbivores.

Aboriginal burning created and maintained dry, infertile soils with clay subsoils. Hard-leaved, fire-dependent eucalypts are adapted to these soils.[5] The trees slowly turn over mature leaves that are, unpalatable and lacking in nutrients. However, when fires scorch their crowns, the old leaves are replaced by new epicormic leaves that are palatable and nutritious. Arbivorous insects proliferate. At Lake Boga the insects produced a bounty of manna for the Aborigines who scorched the mallees.[6] But without frequent, mild fires, soils changed. Mulch built up and nitrogen, instead of being released to the atmosphere by fire, accumulated in the soil. The trees, which were adapted to dry infertile soils, got sick. They started turning over leaves faster, so their crowns were full of softer and more nutritious immature leaves. This was a boon for arbivores which defoliated the trees. The trees responded by sprouting epicormic leaves which were even juicier. Sustained irruptions of arbivores made the trees even sicker, and a vicious circle of chronic decline was initiated.[7]

Aboriginal burning killed most woody seedlings most of the time. It kept woodlands and forests open, grassy, sunny and windy, conditions that favoured frequent mild burning. Burning was the fundamental human economic activity that shaped Australia's environment – maintained the balance of nature – for forty thousand years. Howitt observed that disruption of Aboriginal burning was responsible for rapid woody thickening and shrub encroachment, megafires and chronic eucalypt decline. Mitchell had earlier described the flow-on effects of woody thickening on biodiversity:

> The omission of the annual periodical burning by natives, of the grass and young saplings, has already produced in the open forest lands nearest to Sydney, thick forests of young trees, where, formerly, a man might gallop without impediment, and see whole miles before him. Kangaroos are no longer to be seen there; the grass is choked by underwood; neither are there natives to burn the grass.[8]

In 1831, Protector of Aborigines in Tasmania, G.A. Robinson saw that woody thickening had occurred in the north-east after Aborigines succumbed to an epidemic of influenza.[9] Disruption of Aboriginal burning by European diseases or European settlers allowed woody seedlings to escape their natural controls. Nutrient cycling, hydrology, microclimate, soil properties, and competitive interactions amongst plants and animals were turned upside down.[10] Litter accumulated on the ground, shade increased, air circulation was reduced, established trees got sick and their pests, parasites and diseases flourished together with animals that preyed on the pests. Animals that rely on grasses and open space declined whilst a few species such as cockroaches and bush rats that specialise in damp, shady areas took over.[11]

Following European settlement of the western slopes and plains of New South Wales, open woodlands were heavily grazed until stock were removed or died during droughts. Subsequent rains or floods germinated 'wheatfields' of woody seedlings and there were no Aboriginal fires or European flocks to destroy them. Dense scrubs established and shade tolerant cypress and sheoaks (casuarinas) gained supremacy over intolerant eucalypts. District Surveyor C.F. Bolton in the Central Lachlan-Murrumbidgee District, observed cypress 'pine' scrubs springing up after the 1866 drought.[12] Many areas were subsequently thinned to promote grass and/or timber, and forests including the Pilliga Scrub were born.[13] In western New South Wales, lessees contracted Chinese labourers to ringbark cypress for ninepence per acre. The District Forester at Nyngan reported that "of all the complaints made by white workers against Chinese labour, none ever contained the suggestion that Australians were being deprived of the opportunity of ringbarking, scrubbing or suckering".[14] The suppressive effect of scrub on biodiversity is still evident in firebreaks about twenty metres wide that were left untreated, along the boundaries of many leases. There is no fuel on the ground because nothing grows under the canopy.

Without Aborigines to burn them, seedlings germinating after floods turned open red gum woodlands into scrubs. By 1895 country formerly described as "good forest land" had become impenetrable scrub with bare ground.[15] The scrubs were thinned, and red gum forests were born. Ironically the cypress and red gum forests born of European culture have lately been reserved to protect our 'natural heritage' and they are reverting to dense scrubs and biological deserts. 'Protection' turns the naturally rich mosaics maintained by human economy, into relatively homogenous scrubs of three dimensionally continuous fuels which cannot be burnt by controlled fire under mild conditions. These promote firestorms that sustain megafires when inevitably ignited under extreme conditions. For example, in January 2013 a fire burnt 56,000 hectares of scrub around the Warrumbungles destroying 56 homes and 100 outbuildings, and decimating the local population of koalas.[16]

River regulation had relatively minor and localised impacts on floodplain vegetation,[17] though one wouldn't think so given current schemes to deprive our food-growers of water from regulated rivers, and send it out to sea. Many ecologists would like to squander our scarcest resource so that it can maintain unnaturally stable populations of waterbirds and unnatural red gum forests, as well as unnatural freshwater wetlands in what should be Australia's largest tidal estuary[18] at the mouth of the Murray.

A reader of modern environmental literature would be excused for thinking that the arrival of Aborigines had no impact on the Australian environment and that European settlers changed wilderness to rangelands for their cattle using the axe, the tinderbox and the plough. In 1966, celebrated pioneering green, A.J. Marshall, wrote *The Great Extermination – A guide to Anglo-Australian Cupidity Wickedness & Waste*. He claimed that:

> It was the emergent skin hunters and timber traders who would be the first to do real damage. ... Soon, however was to come the Great Extermination. For this, the sheep farmer is almost entirely responsible ... In order to get more sheep to the acre two things had to be done. ... First, Dad and Dave (and often Mum and Mabel too) set about ring-barking every tree on 'the place'. ... Secondly, the thing to do was kill all the bigger native animals on 'the place'.[19]

When I studied forest science at university in the early seventies, *Environmental Conservation* by Raymond F. Dasmann was a prescribed text. In this book, the famous North American environmentalist wrote that:

> In Australia, the Old Stone Age cultures of an isolated primitive people had endured over thousands of years. The early impact of man in this region had been essentially benign. ... Australia, most

secure in its isolation, in some ways was most drastically affected
by the 'fatal impact' of the western world. … introduced livestock,
logging, uncontrolled fire, and escaped rabbits played havoc with
Australia's vegetation and animal life.[20]

Marshall's and Dasmann's claims contrast rather sharply against the
observations of the sheep farmer Curr. He considered that "it may
perhaps be doubted whether any section of the human race has exercised
a greater influence on the physical condition of any large portion of the
globe than the wandering savages of Australia". Curr made the point
that: "this country possessed from the first, over a great portion of its
area, the inestimable advantage of being ready for immediate use without
the outlay of a sixpence".[21] Curr's *Recollections* also made it clear that the
larger native animals (kangaroos and emus) proliferated as a consequence
of pastoral development.

Despite significant clearing in some regions such as the wheatbelts
in the east and the south-west, the Strzelecki Ranges and the Atherton
Tablelands, Australia retains nearly 90% of its pre-European area
of woodlands and forests.[22] Greens' obsession with trees and forests
is remarkable given that forests comprised less than 10% of the pre-
European landscape and that there have been no extinctions or loss of
biodiversity as a consequence of forest management. In 1966, CSIRO
Scientist J.H. Calaby wrote of the upper Clarence and Richmond Rivers
in northern New South Wales:

> The mammal fauna is the richest in species so far reported from
> any area of comparable size in Australia. … It is of considerable
> significance to mammal conservation in Australia that such a rich
> and varied fauna can still be found in an area with a long history
> of European economic exploitation, and there is no doubt that
> this situation is largely accounted for by the combination of State
> Forests and the local form of management of private and leasehold
> land for beef production.[23]

The benign nature of native forestry operations in Australia has
apparently driven high profile naturists, Professors David Lindenmayer
and William Laurance, to publish a convoluted explanation of why it is

"almost impossible to conduct a 'perfect' study to fully quantify logging effects on forest biodiversity".[24] Presumably a 'perfect' study would show that logging extinguished a species. Ironically, Lindenmayer and Laurance accused forest managers through history, of hubris in the debate over logging "based on our combined experience of 60 years in working in … forests". During 40 years I have noticed that forest workers generally have a better appreciation of practical conservation than academics who visit forests to gather evidence for their hypotheses.

At the time of European settlement, 85% of New South Wales' mammals that were destined for extinction, lived in the Western Division. Forty percent of mammals that lived there were extinguished. But the Western Division was virtually untouched by clearing or logging, and the extinct mammals were medium size 10 g – 5 kg eaters of seeds, herbs and grasses. The extinctions mostly occurred before the overgrazing, rabbit plagues and fox plagues that are typically blamed. Dan Lunney argued unconvincingly that competition by sheep was the ultimate cause, however sheep numbers continued to rise long after the purported date of most extinctions,[25] indicating that there was still considerable herbage available. The extinct animals were attuned to natural cycles of boom and bust, but disruption of Aboriginal burning favoured a few robust native shrubs and grasses over the great variety of more delicate herbs and shrubs that had sustained these native animals over many millennia. Loss of special food resources was later compounded when overgrazing by domestic stock and feral rabbits during droughts, as well as spread of native and exotic weeds, impacted on the remaining native herbs and grasses.

Curr observed that:

> The pigs'-face, once general in that country has also disappeared, a luxuriant growth of grass having taken its place. The same may be said of the cotton bush and other plants. Then again, throughout the continent the most nutritious grasses were originally the most common ; but in consequence of constant over-stocking and scourging the pastures, these, where not eradicated, have very much decreased, their places being taken by inferior sorts and

weeds introduced from Europe and Africa. Thus the capacity of
our pasturage has increased, whilst its quality has decidedly been
lowered.[26]

Modern ecological studies have confirmed that lack of burning or
lack of seasonal grazing by domestic stock, favour a few robust native
and exotic plants allowing them to shade out more delicate native herbs
and grasses.[27] These impacts are exacerbated by fertiliser applications
that favour exotics over natives.[28]

The problem of lack of mild fire was compounded in the late-
nineteenth century because fine fuels were virtually eliminated by sheep,
rabbits and drought, and some heavier soils lost their upper horizons
to wind erosion. Exposure of hard subsoils prevented regeneration of
herbage and promoted dense scrub development. Mild fires could no
longer propagate. In 1997 James Noble produced an interesting history
of the western scrub problem, and research into potential remedies,
called *The Delicate and Noxious Scrub*.[29] This book highlights the difficulties
of restoring functional ecosystems once natural processes have been
disrupted, but it does give some examples of success. Notably, frequent
burning in some shrub-encroached semiarid rangelands can restore the
natural pattern of mallee on dunes and grassland in swales.

In Australia's Western Deserts, Aborigines maintained traditional
mosaic burning practices until quite recently. Medium-sized ground
dwelling mammals persisted along with feral cats under these fire regimes,
and Aborigines incorporated feral cats in their diet. A few intrepid
modern ecologists such as Neil Burrows and Andrew Burbridge have
camped with traditional Aborigines, observing their use of fire to "clean
up country". Since the last desert tribes let go their traditional way of life,
many have reported that animals such as the mitika (burrowing bettong),
mala (rufous hare wallaby) and the golden bandicoot have "gone to
the sky" because of changed fire regimes.[30] One elder, Warlimpirringa
Tjapaltjarri, first made contact with whitefellas as recently as 1984.[31]

Burrows and colleagues documented changes in fire regimes after
Aborigines left the deserts by comparing fire patterns visible on historic
aerial photos against those in recent satellite images. They also gathered

information on newly extinct desert animals by examining and discussing museum specimens with desert Aborigines. Burrows confirmed that, even in the Western Deserts, Aborigines were able to maintain a diverse, fine-grained mosaic of vegetation supporting animals that depend on human fire.[32]

It is clear that human fire was the major force in Australian ecology and Howitt was right to conclude that anything lessening that force would "very materially alter the balance of nature".[33] A second important force for change was alluded to by Curr when he boasted of his success as a squatter:

> in addition to the thirty thousand sheep which I left on the ground in 1851, the run secured was of first-class quality as squatting country, and capable, with the help of a few tanks, of depasturing a hundred thousand sheep.[34]

It was the provision of permanent water for domestic stock by means of dams, ground tanks[VI] and bores that allowed massive overgrazing by domestic stock, kangaroos and rabbits which compounded the problems caused by disruption of Aboriginal burning. A lesser force for change, because of the lesser proportion of forests compared to savannas, was European style forestry. In the early-twentieth century, newly established land management agencies were headed by foresters trained in Europe. They attempted broadscale fire suppression. Megafires and pestilence were the inevitable result. By the mid-twentieth century this once again became blindingly obvious. Broadscale burning including aerial ignition was adopted, and forest health and fire safety improved.[35]

Unfortunately, improved management was short-lived. By 1973 "environmental philosophers" Routley and Routley[36] had established a new religion of tree huggers in Australian academia. Environmental guru, R.F. Dasmann was moved to comment that "Australia's conservation forces were late in evolving, but they are more soundly based in ecological knowledge than were their earlier counterparts in North America or

---

VI Excavated tanks receiving overland flow from rainfall concentrated by shallow banks constructed at a very slight and constant grade over long distances.

Europe".[37] This kind of attitude brought on the dismemberment of ecological science based on observation, thinking and testing. It created a Frankenstein's monster of pseudoscientific consensus that we see today in many academic institutions, societies and journals as well as environmental bureaucracies and NGOs. Since then, the political ascendancy of wilderness warriors has driven unprecedented environmental and socioeconomic deterioration in Australia.

Endnotes

1 Howitt 1891.

2 Mitchell 1848, Curr 1883.

3 Curr 1883, Howitt 1891, Campbell and Hadlington 1967, Martin 1985, Reid and Yan 2000.

4 Curr 1883.

5 McIntosh *et al*. 2005.

6 Curr 1883.

7 Jurskis 2005, Turner *et al*. 2008, Jurskis *et al*. 2011.

8 Mitchell 1848.

9 Flood 2006.

10 Jurskis 2005, 2009, 2011a.

11 Jurskis *et al* 2003.

12 Curby 1997.

13 Rolls 1981, Jurskis 2009.

14 Noble 1997.

15 Manton 1895.

16 General Purpose Standing Committee No. 5 2015.

17 Chesterfield 1986.

18 Marohasy 2012.

19 Marshall 1966.

20 Dasmann 1972.

21 op. cit.

22 AUSLIG 1990, p. 55.

23 Calaby 1966.

24 Lindenmayer and Laurance 2012.

25 Lunney 2001.

26 op. cit.

27 Price and Morgan 2008, 2009, 2010; Prober *et al.* 2002a.

28 Prober *et al.* 2002b.

29 op. cit.

30 Burrows *et al.* 2006.

31 Toohey 2004.

32 op. cit.

33 op. cit.

34 op. cit.

35 Jurskis *et al.* 2003.

36 Routley and Routley 1973.

37 op. cit.

5

IF THERE IS NO BALANCE OF NATURE, HOW CAN IT BE DISTURBED?

That is the one point which I think all evolutionists are agreed upon, that it is virtually impossible to do a better job than an organism is doing in its own environment.

Richard Lewontin

In my view and, I think, Charles Darwin's,[1] natural succession is a process that occurs on geological time scales, and is one and the same with evolution. The more popular modern concept of succession seems to be that biota establish on extensive new substrates such as volcanic ash or dunes of raw sand, and successively give way to other biota in a process analogous to the development of an individual organism towards maturity. This concept is arguably a precursor to the modern religion of Gaia. The reputed founder of ecology, Danish botanist Eugen Warming introduced it in 1895[2] and four years later American ecologist Henry Chandler Cowles popularised it in English after studying vegetation chronosequences[VII] on dunes at Lake Michigan.[3]

I vaguely remember some schoolteacher, in some classroom, explaining succession in something like the following terms: A bare rock surface is too hard and dry to grow trees. Lichens, or maybe mosses, establish and start chemically breaking the rock surface and trapping dust and moisture so that a thin soil gradually develops. Over time a succession of bigger plants creates a deeper and moister and more nutritious soil, and helps to break up the rock with their roots until eventually trees can grow. As bigger bushes and trees establish, they cast increasing shade so

---

VII Spatial sequences of different age classes of vegetation on progressively developing substrates such as sand dunes.

that shade-loving plants can establish. At the same time, light-demanding 'pioneer species' start to die out. As more shade and shelter develop, various epiphytes establish on trees. Eventually the process is complete and there is a climax rainforest. I didn't wonder at the time whence came the bare rocks and the various propagules. Nor did I wonder why there are some rocks that are still bare after thousands of years.

Ironically the strange idea of secondary succession, appears to have pre-dated the logically sound, though empirically difficult, concept of primary succession. Adolphe Dureau de la Malle is credited with the first use of the term "succession" in an 1825 paper on vegetation development following clearfelling of forest.[4] In 1860 Thoreau gave an address to a gathering of the Middlesex (a County in Massachusetts) Agricultural Society. This annual gathering was known as *Cattle-show*. The title of Thoreau's address was *The Succession of Forest Trees*[5] wherein he attempted to explain "how it happened, that when a pine wood was cut down an oak one commonly sprang up, and vice versa". Thoreau discussed differences between oak and pine in seed dispersal mechanisms, requirements for germination and establishment of seedlings, and longevity of seeds. For example pine seeds are dispersed by wind and cannot germinate in heavy shade or on top of a thick layer of needles, whereas acorns and other nuts are dispersed and 'planted' by squirrels and mice, and can germinate successfully under shade and a mulch of fallen leaves.

Frederic Clements published papers in 1904 and 1905 drawing a distinction between primary succession as described by Cowles and secondary succession, as discussed by Thoreau, where disturbance is said to restart a predictable progression towards a stable climax community. His magnum opus, *Plant Succession*, was published in 1916[6] and dominated ecological thinking until late in the century.

But in 1911 Cowles had written: "This classification [of primary versus secondary succession] seems not to be of fundamental value, since it separates such closely related phenomena as those of erosion and deposition, and it places together such unlike things as human agencies and the subsidence of land".[7] However, instead of recognising that primary succession is one with evolution, pinning down the nature of disturbance, and throwing out the illogical concept of secondary succession, Cowles' contemporaries continued to argue about the pathways by which succession progressed.

Empirical difficulties with the Clementsian theory of disturbance and succession were such that Henry Gleason published a paper in 1926 called "The Individualistic Concept of the Plant Association".[8] He argued that individual species prospered or not according to an environmental luck-of-the-draw and that so-called plant associations were a coincidental outcome. Gleason's concept is supported by recent evidence of vegetation development after volcanic eruptions, where plants establish according to the availability of propagules of species adapted to grow in ash or in cracks in solidified lava flows.

In 1966 Leslie Viereck published a study of primary succession on gravel outwash from a receding glacier in Alaska.[9] His introduction gives a plausible account of a series of terraces of different ages supporting a progression of vegetation over several centuries from a pioneer stage, through a meadow stage, an early shrub stage, and a late shrub stage, and finally reaching a "climax" tundra after about 7000 years. Plausibility is diminished as one reads the body of the report and the discussion. It seems that shrubs of willow, poplar, birch and spruce are scattered through the different successional stages. More importantly, there is a gap of seven thousand years between these stages and the climax tundra whose parent material "is thought to be" outwash from a different glacier.

This study could possibly be seen as supporting a hybrid model between Clementsian succession and Gleasonian luck-of-the-draw, at least over a few centuries up to the late shrub stage. As such it is more satisfying than the Clementsian concept of orderly progression because it marries the empirically sound concepts of soil development over time and haphazard establishment opportunities. Unfortunately it doesn't address the major problem that there are no chronosequences extending from the creation of new substrates through geological time to complex highly developed ecosystems. Only a chronosequence of planets would address this problem.

Australian ecologists Connell and Slatyer have been credited with developing better models of succession that have allowed ecologists to move on from the Clementsian view.[10] They proposed three models called Facilitation, Tolerance and Inhibition, but found little evidence for the Tolerance Model.[11] Their Facilitation Model seems essentially the same as Clements' model of primary succession, where pioneer species create suitable conditions for later successional species. The Inhibition Model or 'first in best dressed' seems to me a restatement of Gleason's concept.

I have no problem with primary succession except for the time scale and the fact that it is an historical model rather than a predictive one. A balance of nature evolves from the interaction amongst biota and their environment over geological time. Primary succession can be characterised as the long-term competitive process of evolution in response to geological processes that create new environments. Thus primary succession began with life on earth about three and a half billion years ago. It is the process that Darwin called Natural Selection and Herbert Spencer called Survival of the Fittest.

Natural speciation and extinction proceed at an incredibly slow pace. In practice, on human time frames, new environments very rarely arise. For example, raw sand dunes may extend on one side whilst they contract on the other as soil and vegetation develop. Dunes, lava flows, newly exposed rock surfaces and the like are quickly colonised by adapted biota within dispersal distance. Biota that are adapted to each stage of soil/vegetation development merely shift across the landscape.

The eruption of Mt. St. Helens in 1980 provided a great opportunity to test hypotheses about disturbance and succession. Landslides and pyroclastic flows (superhot gas and rock) obliterated 60 square kilometres of forest. Another 500 square kilometres of vegetation was blown down or scorched, and a further 1000 square kilometres was blanketed with tephra (fine rock fragments) to a depth of five centimetres.[12] This was a natural event by any definition, and it seemingly created new substrates. United States' Congress established a 43,000 hectare National Volcanic

Monument where scientists studied 'succession' for quarter of a century before declaring that they had learnt:

> important lessons. First, living and dead biological legacies (for example, dead trees and rotten logs) are integral to the ecological response, even after severe disturbances. Second, ecological succession is very complex, proceeding at varying paces along diverse paths, and with periodic setbacks through secondary disturbances. Consequently, no single, overarching succession theory provides an adequate framework to explain ecological change.[13]

Prior to 1980, Mt. St. Helens volcano was active over eight different periods during the 16,000 years that humans have occupied North America.[14] It was active from 1800 to 1857 and from 1479 to 1720. Before that there was one period of activity lasting less than a century. Five other active periods lasted several centuries. My colleague John Turner tells me that two other recently active volcanos, Mt. Rainier and Mt. Hood, are visible from the top of Mt. St. Helens, and a third, Mt. Mazama, is no longer visible because its last eruption turned it into Crater Lake. John was puzzled by a strange 'bleached' layer in nearby soils until he discovered that it was ash from the eruption. I would suggest that volcanic activity is a natural phenomenon which all the organisms and ecosystems within this region have experienced and to which they are attuned. To my mind it is not a disturbance.

There are organisms in the region that are suited to all the 'new' substrates or opportunities created by volcanic eruptions, and to the "living or dead biological legacies (for example, dead trees and rotten logs)" that afterwards remain. The ecological response seems quite simple to me. Each 'new' substrate is occupied or used by those plants and animals that have the ability and the access. Gleason's concept of luck-of-the-draw is not a theory of succession so much as an argument against the idea of secondary succession. It is a statement of fact. This is not a "very complex" process, but it produces a complex pattern in the landscape because a variety of physical changes arising from volcanic eruptions are superimposed on an already variable landscape. A theory of succession is not necessary to explain this simple process of species shifting across a complex and changing landscape.

John Turner is somewhat an amateur vulcanecologist. He informs me that volcanos in Hawaii produce lava flows rather than ash and tephra because they contain less viscous, basic magmas that are less likely to explode than more viscous acidic magmas in the Pacific north-west of the continental United States. John says that new lava flows in Hawaii are typically colonised by the naturalised Australian tree, silky oak (*Grevillea robusta*), whose winged seed is dispersed by wind, and is able to germinate and establish in the deep cracks which typically occur as lava flows cool and solidify. 'Natural experiments' such as this show that Gleason's ideas were 'spot on'.

In a modern Australian textbook, Peter Attiwill and Barbara Wilson discussed what they saw as a primary succession on the Cooloola Sands. They referred to a study by Walker and colleagues[15] of a progression of species across the landscape, but in this case the progression was apparently confounded by a true disturbance – exclusion of mild fire. In the natural scheme, as soon as enough flammable fuel had accumulated on the originally raw sand to carry a fire, Aborigines would apply the firestick. In the absence of the firestick, by the time blackbutt (*E. pilularis*) woodland had developed at the back of the dunes, nitrogen was accumulating at an unnaturally high rate, soil acidity and moisture content were unnaturally high and the roots of the blackbutt trees were deteriorating. The trees were not adapted to an environment that lacked sustaining fire.[16]

The textbook described this progression of species across the landscape, ending in chronic decline of the blackbutt trees, as a primary succession that is initially progressive and later regressive as excess nutrients are released to the soil and lost through leaching. In fact, it is an example of a global phenomenon that occurs when nitrogen accumulates in soil as a result of true disturbance such as fire suppression, atmospheric pollution, urban runoff, fertiliser application and so on. A feature article published by the Ecological Society of America in 2010 explained that "N accumulation is the main driver of changes to species composition across the whole range of different ecosystem types" around the world.[17]

Attiwill and Wilson stated that "With periodic, natural disturbance by fire, regressive succession does not appear to be common in most Australian communities, but it is obvious in some communities where fires have been suppressed".[18] However, it is quite clear that frequent mild fire is a natural component of most Australian ecosystems and fire suppression is a disturbance that upsets the balance of nature. An unintentional 'experiment' on the south coast of New South Wales has clarified the natural process on sand dunes. Long-unburnt blackbutt woodlands adjacent to Merimbula Airport are in chronic decline (Fig. 4) whilst frequently burnt blackbutt woodlands adjacent to Moruya Airport are healthy.

'Secondary succession' can best be seen as a short-term competition amongst biota whose interactions are well established. If their physical environment has not been fundamentally altered, their interactions should be entirely predictable. Thoreau was on the right track in pointing

Fig. 4 Long-unburnt, chronically declining blackbutt woodland invaded by pittosporum on rear dunes at Merimbula. (Photo by author)

this out but he should also have considered why oak woods existed at all before European settlers started felling pine woods. Ecological history is the key to this dilemma. Burning by Native Americans gave oaks a competitive advantage over pine just as burning by Australian Aborigines gave lignotuberous eucalypts a competitive advantage over cypress on the western plains of New South Wales.[19] Pine woods developed after Native American burning was disrupted, but there was no predictable progression towards a stable climax. Oak and pine had reached a balance maintained by human fire over about 16,000 years. European settlers fundamentally altered the environment by disrupting burning. They disturbed the balance of nature. Under the Healthy Forests Restoration Act of 2003, conservation managers in USA are increasingly using thinning and burning to restore the balance.[20]

When I discussed a draft of this chapter with my mentor and friend, Peter Attiwill, I was surprised that he thought my treatment superficial, and also at the vehemence of his reaction to Clementsian thinking. I attach no more significance to the wrong thinking behind Clementsian succession than to that underpinning the replacement concepts of "state and transition", "multiple stable states" or "novel ecosystems". But Peter regards Clementsian Succession as more a religion than a theory. He incanted some of the special words associated with it, such as "climax", "pristine" and "harmony", and voiced his opinion that "the damage that Clements did to independent thinking over much of the twentieth century was a disaster".

Although I empathise with Peter's reaction to the religious words and thoughts that Clements shared with modern 'wilderness warriors', I don't see that Clements did **exceptional** damage to independent thinking. I see environmental disasters such as megafires and chronic tree decline still happening in the twenty-first century because editors of scientific journals publish completely wrong thinking that conforms to 'the consensus', and suppress independent thinking because it is independent. Wilderness warriors publicise this pseudoscientific groupthink to influence politics. To my mind, the main issues are that many ecologists try to separate man from nature, that they hold a concept of natural disturbance and that they confuse a natural process that occurs over geological time with the havoc wreaked by true disturbance over human time frames. The new concepts that have been proposed to replace the failed concept of secondary succession cannot make order out of this chaos any more than did the original.

Having, to my mind, 'nailed' the issues identified by Peter, I sent my manuscript to ecological historian Charles E. Kay at Utah State University for comment. Charles' work emphasises the role of Aboriginal people in pre-European ecosystems, but his reaction to these issues was even more vehement than Peter's. He advised me that "the balance of nature" is a very old, religious, loaded term that should never be used, and pointed me to two books by Daniel Botkin that supposedly debunk this ancient myth. Kay suggested that my concept of a dynamic balance of nature is oxymoronic.

I read Botkin's recent work *The Moon in the Nautilus Shell. Discordant Harmonies Reconsidered*, and I found that he also equates balance with stasis. At the start of the book, Botkin uses Venice, founded in the fifth century AD, as an example of constant environmental changes and man's problematic interactions with them: "tidal action … has kept Venice livable throughout the centuries. … Venice has therefore depended on environmental change on a twice-daily basis, and … tides are a global phenomenon".[21]

Obviously, dynamism is not imbalance, otherwise harmonic balancers wouldn't prevent crankshaft failures, tightrope walkers would never reach the far end and gyroscopes would be of no practical use. Tides, seasons, droughts and floods are part of nature, they are not changes or disturbances. Rather than looking to Venice as a model of environmental problems through 1,500 years, we can look to Aboriginal Australia as a model of solutions spanning 50,000 years.

Australian Aborigines originally disturbed a dynamic balance of nature. The firestick replaced a regime of infrequent high-intensity lightning fires. They rearranged the vegetation and extinguished many plants along with the megafauna that ate them, before maintaining a newly established balance through forty thousand years of large environmental fluctuations. Aborigines persisted through droughts, floods, a glacial maximum, volcanic eruptions and hugely rising sea levels. I agree with Botkin, once disturbance has occurred through human arrival in a former wilderness, "that human actions are required to create a balance". Australian Aborigines disturbed a wilderness and proved that there can be a balance of nature including man. Mitchell saw this dynamic balance; "Fire, grass, kangaroos, and human inhabitants, seem all dependent on each other for existence in Australia".[22] Unfortunately European explorers didn't recognise a similar balance in North America before it was disrupted.

A good example of the confusion arising from a concept of natural disturbance and succession appeared in the prestigious *Tansley Review* series in 2012. The authors proposed that giant eucalypts such as mountain ash (*E. regnans*) or flooded gum (*E. grandis*) growing over rainforest subcanopies should be regarded as pioneer rainforest species.[23] In fact these wet eucalypt stands are mostly remnants of more extensive eucalypt stands that were replaced by rainforests as climate ameliorated after the last glacial period.[24] They survive because they are accessible to lightning fires or ember attack from nearby eucalypt forests during extreme weather conditions. Such fires occurred naturally at intervals of a few centuries.[25] Eucalypts and rainforest trees regenerated simultaneously after these fires and their relative numbers were maintained through the centuries.[26] There was neither disturbance nor succession.

In 1930 Charles Elton, perhaps unintentionally, focused attention on the nature of disturbance rather than pathways of succession by rejecting the ancient wisdom of a balance of nature. In *Animal Ecology and Evolution* he wrote:

> At the same time it is assumed that an undisturbed animal community lives in a certain harmony referred to as the balance of nature, and that although rhythmical changes may take place in this balance, yet that these are regular and essentially predictable and, above all, nicely fitted into the environmental stresses.... The picture has the advantage of being an intelligible and apparently

logical result of natural selection in producing the best possible world for each species. It has the disadvantage of being untrue. The balance of nature does not exist and perhaps never has existed.[27]

In 1983 Connell and Sousa wrote "if it does exist it is exceedingly difficult to demonstrate".[28] As evident from Botkin's books, the concept of a balance of nature remains unfashionable. Daniel Simberloff wrote:

> For academic ecologists, the notion of a balance of nature has become passé, and the term is widely recognised as a panchreston … a term that means so many different things to different people that it is useless as a theoretical framework or explanatory device.[29]

But ecologists cling to the notion of disturbance, and difficulties continue to arise because they have been generally unable to come up with a useful way to define it. Having found that the idea of secondary succession doesn't work, they've replaced it with equally unworkable concepts. These beg the question – what difference between state and balance? They imply that some sort of balance or stability has been disrupted and a new balance established.

I'm grateful to Peter Attiwill for prompting me to dig deeper into the literature around disturbance and succession. I was surprised to find that at least one prominent ecologist still thought in terms of a balance of nature as late as 1991. Stuart L. Pimm wrote: "What is this balance of nature … If ecologists now speak more of ecological stability, perhaps we are just substituting one phrase for another. *Balance* or *stability* implies some restoration following disturbance".[30] Some ecologists like Botkin and Simberloff have attempted to resolve the problem of "disturbance" by rejecting the notion of "balance", whilst others have made "disturbance" all-encompassing, and therefore meaningless. In 1985 Pickett and White published The Ecology of Natural Disturbance and Patch Dynamics, wherein they stated that:

> we define "disturbance" not in a sense that is relative to the "normal" environment but rather a more tractable and physical

sense. Thus, our definition of "disturbance" includes environmental fluctuations and destructive events, whether or not these are perceived as "normal" for a particular system.[31]

Elton appears to have been narrowly focused on animals and especially numbers of animals in contrast to early Australian ecologists such as Mitchell who took a broader view of the interaction between plants, animals, fire and man. They recognised climatic variability and irregular cycles of boom and bust. Animal populations are constantly fluctuating because the size and extent of food resources, starting with plants and moving through plant-eaters and predators of plant-eaters, fluctuates with climate.[32] The early Australian naturalists saw that man was able to influence his food resources using the firestick. Thus they were able to appreciate a dynamic balance of nature including man.

Australia's prehistory and history indicate that human arrival was a disturbance that disrupted the previous balance of nature and extinguished the megafauna. Over a few millennia, a new balance including human fire was established. The balance fluctuated spatially according to climatic fluctuations. For example, the sharp boundary between rainforests and eucalypt woodlands on the Atherton Tableland shifted inwards and outwards[33]. In the Bega and Towamba valleys, snow gum (*E. pauciflora*) retreated upwards at higher altitudes and downwards into frost hollows at lower altitudes as climate ameliorated, so that there is currently a disjunct distribution with a few isolated low altitude occurrences. European settlement was another disturbance that upset the balance created by Aborigines.

Disturbance needs to be tightly defined. Droughts and floods have, until the recent advent of climate hysteria, been regarded as natural disturbances whilst burning by humans has been seen as anthropogenic disturbance. But eucalypt ecosystems were shaped by droughts and/or floods as well as human burning over many thousands of years. These factors maintain a dynamic stability by controlling plants and animals that are not adapted to them whilst favoring biota that are adapted and, indeed, dependent on them. Disturbance should be seen as an unnatural factor outside the experience of the ecosystem in question. To define disturbance, it is essential to examine historical evidence as well as physical evidence of the environmental history of the ecosystem.

After 380 pages of discussion involving 31 ecologists, Pickett and White concluded that "Although an understanding of disturbance is of crucial importance in ecology, no coherent theory exists to further its study".[34] There is no coherent theory of disturbance because there has been no coherent definition of disturbance. Pickett and White were searching for a theory that encompasses natural events with predictable outcomes as well as unnatural events with unpredictable outcomes. Their book was transferred to digital printing in 2005 and was still in print in 2014. Confusion continues because ecologists have failed to recognise that natural events are not disturbances. Only new events originating from outside the natural environment are disturbances. No theory of succession can predict the outcomes of disturbance. Succession is an historic model not a predictive one.

The arrival of the Aborigine in Australia with his firestick was a disturbance that initiated a succession which extinguished some soft-leaved flora as well as the megafauna that depended on them. After many generations Aborigines had established a new balance that persisted for about forty millennia. Aborigines always carried fire whether on land or water, and used it for heating, cooking, hunting, signaling, fighting, celebrating, managing vegetation and for fun.[35] Aboriginal fires felled trees and consumed fallen timber, litter, seedlings, ground layer vegetation, young shrubs, saplings and animals' nests.[36] They kept wet sclerophyll forests, scrubs, rainforests, rotten logs and animals that depended on these features contained in physical refugia. Early naturalists such as Tench[37] recognised that some highly mobile animals ranged across savannas and retreated into refugia at night or when threatened by predators or fires.

Frequent low-intensity fires in the landscape maintained nutrient cycling that supported mostly healthy trees or shrubs with an advantage over their pests, parasites, diseases and competitors including seedlings of their own kind.[38] Infrequent moderate to high-intensity fires affected refugia within this matrix individually because others were isolated from ignition by the reduced flammability of the frequently burnt matrix.[39]

The small, isolated, high-intensity fires did not initiate any succession because the species occupying the refugia were triggered to regenerate and the fires merely reset the stands to a young age.[40] In the wet tropics, Aboriginal fires drew a sharp line between rainforests and eucalypt woodlands.[41] In the spinifex and the mallee, Aborigines burnt in mosaics that reinforced the physical mosaic of dunes and swales.[42] Fires were integral to the balance of nature, they were not a disturbance.

Similarly, floods maintained a balance on the floodplains by recharging the aquifers that sustained river red gum trees and drowning other vegetation that established between floods. When a river red gum reached the term of its natural life (probably half a millennium or more) it left a large root mound that would provide a suitable site somewhere within its range of elevation for seedlings to establish after the next flood whatever the height of the flood.[43] Droughts maintained a balance on higher flood plains by favouring drought-tolerant and flood-tolerant black box [*E. largiflorens*] trees and killing other vegetation that established during moist seasons. Emus, kangaroos and waterbirds avoided droughts by migration, or stopped breeding and persisted in lower numbers. On sandhills or rocky hills in arid zones, droughts favoured extremely drought-tolerant trees such as cypress and mallees.

Fire, droughts and floods stressed weaker trees more than stronger trees and this allowed pests, parasites and diseases to overcome these trees' weakened defenses. The arbivores obtained enough food to grow to maturity and breed. The balance was dynamic but the arbivores did not cause decline of fundamentally healthy trees. The struggle for existence ensured that fittest trees survived as did their fittest predators. Fires, droughts and floods maintained a dynamic, resilient balance of nature.

The arrival of dingoes in mainland Australia was a disturbance, but only a minor one. As Mitchell deduced, they displaced two other species with similar diets. Introductions of exotic fauna by European settlers were also minor disturbances. The fox and the cat hastened the demise of some fauna that were disadvantaged by vegetation change, whilst the rabbit substituted for Aboriginal fire and controlled establishment of woody seedlings until myxomatosis was introduced to control rabbits.

When Europeans stole the firestick it was a major disturbance, except where seasonal grazing by domestic stock substituted for burning and maintained ecosystem function. Grazing killed most woody seedlings and maintained grassy open forests and savannas. After stock perished in droughts, drought-breaking rains or floods germinated millions of woody seedlings that grew into scrubs. Foresters thinned scrubs of cypress and scrubs of river red gum, restored grass and reintroduced stock.[44] The new forests that they created would fit the modern ecological concept of novel ecosystems except that they depend on ongoing human economy to function. The proponents of 'novel ecosystems' seem to imagine that ecosystems created by human economy can sustain themselves without any further human input. In 2009 an article in *Nature* asserted that: "A novel ecosystem is one that has been heavily influenced by humans but is not under human management. A working tree plantation doesn't qualify; one abandoned decades ago would".[45]

But landscapes that have been abandoned or 'protected' by excluding human economy are certainly not "working", they are in chaos. Former woodlands of mallee, cypress or river red gum that have been 'protected' in national parks now comprise dense young even-aged scrubs. Old trees, herbs and grasses, bare ground and the rare fauna that depend on these elements of habitat are vanishing fast.[46] These are not alternative stable states or novel ecosystems because they cannot sustain themselves. An open woodland near Geelong that was first described by Matthew Flinders in 1802 has been degenerating and losing old trees and biodiversity for two centuries since Aboriginal burning was disrupted. During the second half of the twentieth century, sheoak increased in density from tens of stems per hectare to more than 3000 stems per hectare.[47] A similar process occurred in the iconic bushland of Kings Park in Perth, where wildfires created sheoak scrub from jarrah woodland.[48] These are examples of disturbance and 'secondary succession'. The end points are still unpredictable. Processes such as this are the genesis of the Gleason/ Inhibition model of secondary succession. However it's not a case of 'first in best dressed'.

The problem is that the environment within which competing species have evolved over geological time has been fundamentally changed. Their interactions are no longer predictable because true disturbance disadvantages some species and temporarily favours others. Those that are initially favoured eventually 'hit the wall' because some aspect of their life strategy is not attuned to the new environment. For example, mature trees cannot survive in the sheoak scrub near Geelong. The dense young stands will not be able to regenerate when they are inevitably killed by drought or high-intensity crown fire, because flowering and seeding is suppressed in dense stands. This problem cannot resolve itself over human time scales, but man can resolve it by restoring open conditions and mild fire.

In Western Deserts where Aboriginal fire management remained in place until quite recently, mesofaunal extinctions did not occur despite the arrival of an exotic predator, the feral cat. Rather, a fine adjustment occurred as Aborigines incorporated cats in their diet.[49] Changing a few elements of an ecosystem has minor consequences compared to changing the way an ecosystem functions. The persistence of Aboriginal culture, mesofauna and a landscape scale vegetation mosaic through forty or fifty thousand years of climate change and hugely fluctuating sea levels give testimony to a robust man-made balance of nature. The calamities that ensued after European settlers extinguished the firestick are testaments to the true nature and consequences of disturbance.

Aboriginal burning, lightning strikes, floods, droughts, windstorms, cyclones and volcanic eruptions are natural phenomena. Human arrivals in unpeopled regions were disturbances because the firestick upset the balance of nature and initiated succession over geological time towards a new balance. Secondary succession, alternative stable states and novel ecosystems are not useful concepts because they deny human economy. Once there has been a true disturbance, chaos ensues until such time as human economy can re-establish stability.

Darwin's theory of evolution was grounded in prehistory and fertilised by his observations of the outcomes of deliberate breeding by man of pigeons, poultry, pigs and so on. It was informed by the universal struggle for existence that maintains a balance of nature. To deny a balance of nature is to deny evolution. To deny man's place in nature is to deny history. Modern ecology was cast adrift when it slipped its moorings in prehistory and history. The 'baby' of balance has been thrown out with the 'bathwater' of secondary succession.

Endnotes

1 Darwin 1860.

2 Goodland 1975.

3 Cowles 1899.

4 Dureau de la Malle 1825.

5 Thoreau 1860.

6 Clements 1916.

7 Cowles 1911.

8 Gleason 1926.

9 Viereck 1966.

10 Attiwill and Wilson 2006.

11 Connell and Slatyer 1977.

12 Dale *et al*. 2005.

13 Ibid.

14 Waters *et al*. 2011, U.S. Geological Survey 2005.

15 Walker *et al*. 1981, Attiwill and Wilson 2006.

16 Jurskis 2005, Turner *et al*. 2008, Jurskis *et al*. 2011.

17 Bobbink *et al*. 2010.

18 op. cit.

19 Jurskis 2009.

20 Hessburg *et al*. 2005.

21 Botkin 2012.

22 Mitchell 1848, p. 412.

23 Tng *et al*. 2012.

24 Singh *et al*. 1981.

25 Turner 1984.

26 Jackson 1968, Fig. 2.

27 Elton 1930.

28 Connell and Sousa 1983.

29 Simberloff 2014.

30 Pimm 1991.

31 Pickett and White 1985.

32 White 1993.

33 Singh *et al.* 1981.

34 op. cit.

35 Tench 1793; Sturt 1833; Mitchell 1839, 1848; Curr 1883; Howitt 1891; Burrows *et al.* 2006.

36 Ibid.

37 op. cit.

38 Jurskis 2005, Turner *et al.* 2008, Jurskis *et al.* 2011.

39 Jurskis 2011b.

40 Jackson 1968, 1999.

41 Stanton *et al.* 2014b.

42 Noble 1997, Burrows *et al.* 2006.

43 Jacobs 1955.

44 Jurskis 2009.

45 Marris 2009.

46 Jurskis 2009, 2011a.

47 Lunt 1998.

48 R. Underwood, pers. comm. 2015.

49 Burrows *et al.* 2006.

6

RARITY, DECLINE AND PESTILENCE

Plenty orright, too much no good fucken.
Italian migrants working on Snowy Mountains
Hydroelectric Scheme circa 1950

Darwin considered that rarity was a precursor to extinction because rare species present less variability upon which natural selection – the struggle for existence – can operate.[1] I think this is true on geological time scales but need not apply on human time scales. Species may be naturally rare if they have very specialised habitats, and they may be declining if their habitat is deteriorating or shrinking. Unfortunately environmental legislation and relevant bureaucracies in Australia don't sufficiently recognise this distinction, and don't appreciate the extent to which species and habitats are being lost as mild fire is no longer applied to sustain them.

Mount Imlay mallee (*E. imlayensis*) is a case in point. It is naturally rare because its habitat is skeletal soil on Devonian, Merimbula sediments at high altitude and moderate latitude near the coast. This habitat occurs only atop an isolated mountain near Eden, where there is a population of about 70 plants from probably only five clones.[2] Of itself, this is not a problem. For example, a rare sub-alpine stand of mostly clonal Huon Pine at Mt. Read in Western Tasmania has survived for more than a millennium, enduring both the Little Ice Age (which was moderated by the southern ocean) and the Medieval Warm Period.[3] However the Imlay mallees are declining.

There have been only two fires near the site in over 60 years, neither of which apparently touched the mallees. The trees are nearly all unhealthy and about 10% of the population died during the first decade of the twenty-first century.[4] The species is officially listed as critically endangered, and the conservation authorities consider that it is threatened by drought, fire, insect attack and the root rot fungus *Phytophthora cinnamomi*.[5] They are concerned about unreliable seed production, lack of seedling recruitment, limited genetic diversity, and 'disturbance' by fire. But Stephen Pyne aptly described mallees as pyrophytes.[6] The Imlay mallee is threatened by **lack** of fire. Its habitat is disappearing in scrub.

Botanists studying the mallees can't use a GPS to map them because a dense subcanopy of tea tree interferes with reception of satellite signals.[7] The tea tree scrub is obliterating the naturally dry, sunny and airy habitat of the mallee. As its roots deteriorate, Imlay mallee is becoming susceptible to drought and attack by parasites, predators and diseases – collectively known as arbivores. There is phytophthora at the site, and there are dead stems of mallee. Devil's twine or dodder laurel (*Cassytha*), is strangling the sick trees whilst psyllids attack their leaves. In its debilitated state, this natural pyrophyte is liable to be killed by a fire fierce enough to penetrate the scrub. The mallees are mostly not producing seed because epicormic shoots are repeatedly shed before the long cycle from budding through flowering and pollination to seed maturation can be completed. In any case, there is no suitable seedbed because the ground is covered with moss and litter, whilst sunshine and fresh air are excluded by scrub.

The parks authority's well publicised response to this crisis was to raise a handful of seedlings at the National Botanic Gardens and plant them back into the now unsuitable habitat. Of course there were boot washing facilities to ensure the planters didn't spread phytophthora, but we weren't told how the bush rats and cockroaches were trained in forest hygiene. Planters were ferried up the mountain by helicopter, and the stunt was advertised as an example of active conservation.[8] It would have been a bloody good joke if not for the expense to the taxpayer and the inevitable outcome confronting the endangered species.

Chronic eucalypt decline is an outcome of disturbance to the balance of nature. The disturbance in most cases is fire suppression, but it can be pasture improvement, earthworks, water impoundment or diversion, industrial or urban pollution, or any combination of these. Disturbance changes the physical and chemical properties of the soil around the roots of the tree, and the roots deteriorate.[9] This can happen to a part of the root system, where, for example, a tree 'straddles' a management boundary; a whole root system; or a whole stand. The effect is initially reflected in the decline of the canopy served by the affected roots, whether part or the whole canopy of the tree or stand. As populations of arbivores build up in response to the improved nutrition provided by the sick canopy, damage can extend further through still healthy canopy. Arbivores eating healthy leaves die off because healthy leaves are poor food. But the dead bugs are replaced by offspring of healthy bugs feeding in the sick canopy. Some of these will breed successfully because they will benefit from the soft and juicy epicormic foliage that grows back in the previously healthy canopy after its defoliation by the unlucky outcasts of the previous generation.

Chronic decline is extending through eucalypt ecosystems across Australia. It affects almost 20% of forests and woodlands in coastal New South Wales, whilst more than 50% of eucalypt systems in Queensland's wet tropics are declining.[10] Decline is widespread in Victoria, Tasmania and the south-west of Western Australia, but reliable figures are not available. Forest pathologists are mostly ignorant of the fundamental problem and the simple solution. Australia's State of the Forests report for 2013 reflected their confusion:

> Many pests and diseases, particularly native ones, exhibit cyclical patterns of impact on native forests, and are generally of minor overall concern. ... A wide range of persistent or intermittent crown dieback syndromes occurs to some degree in native forests in all states and territories, often resulting in significant tree mortality and associated ecosystem impacts. These syndromes are usually caused by combinations of factors such as climatic stresses, poor land management practices, severe insect attacks, and an imbalance in insect predator levels; ameliorating their impacts through forest management can be difficult.[11]

One hundred and fifty years ago, Howitt recognised chronic decline of red gum defoliated by insects in East Gippsland as a consequence of not burning.[12] Twenty-five years ago Bob Ellis working in forests,[13] and Jill Landsberg working in woodland pastures,[14] elucidated connections between soil changes, tree physiology and chronic decline. Not burning native pastures, or sowing and fertilising exotic pastures, have similar effects on soils, tree roots, resilience to drought and waterlogging, and food quality of trees for arbivores.[15] From the mid-twentieth century, a government bounty on fertilisers was responsible for an upsurge in rural tree decline.[16] Since the late-twentieth century, expansion of conservation reserves, and reduction of burning in native forests and woodlands has caused a resurgence of forest decline.[17]

Lack of frequent, low-intensity burning allows a mat of litter and woody seedlings to develop, choking out herbs, grasses and bare ground, and affecting microclimate, nutrient cycling, soil conditions and ultimately the health of established trees. Ironically, modern ecological assessments of ecosystem health typically identify absence of this developing problem (i.e., lack of eucalypt saplings, litter, fallen timber and shrubbery) as a sign of degradation. For example Gibbons and colleagues thought that the size class distribution of eucalypts in a healthy system should have a reverse J shape.[18] This is incorrect because the reverse J curve distribution indicates that seedlings and saplings are proliferating at the expense of declining trees. Such stands are proceeding down the vicious spiral of chronic decline for the want of mild fire.

Mulch builds up, sunshine and air circulation are reduced. Nitrogen in litter, seedlings and herbage that had previously been volatilised by fires and returned to the atmosphere, or mineralised by fires and taken up by the flush of new growth, now accumulates in the soil and the developing shrubbery. Topsoils become cooler, damper, softer and deeper. Carbon to Nitrogen ratios of soils are reduced, they become more acid (except in the case of some calcareous soils), and microtoxins such as aluminium and manganese are released. These inhibit tree roots and mycorrhizae. They become more susceptible to droughts and root rots such as phytophthora. The deteriorating soils and roots cause nutrient imbalances and physiological changes in the trees. Their sapstreams and foliage become more attractive and nutritious to arbivores – that is anything that derives nutrients from any part of the tree including roots, sapwood, sap and leaves.[19]

Peter McIntosh and colleagues compared soils from dry eucalypt forests in Tasmania, developed under the influence of frequent burning by Aborigines, with soils from moist eucalypt forests,[VIII] and soils developed on similar substrates in New Zealand without the influence of human fire. Soils under Tasmanian dry forests have less nutrients, higher Carbon to Nitrogen ratios and contrasts in texture with depth. Frequent burning causes clay particles to migrate from topsoils into deeper profiles because there is virtually no organic matter to bind them at the surface. Clay subsoils impede drainage. Soils under moist forests, and New Zealand soils have uniform or gradational textures. Tree roots can more freely access both air and moisture. Frequent burning in dry forests volatises nutrients or releases them to runoff and leaching. McIntosh and colleagues pointed out that removal of nutrients by frequent burning "encourages fire-tolerant vegetation adapted to lower soil nutrient status".[20] Disruption of Aboriginal burning and subsequent fire suppression overturned the natural processes of soil formation and created soils that are no longer 'in tune' with trees that grow on poorly drained infertile sites.

Arbivores include phytophthora, armillaria, various borers and sapsuckers, leaf-eating insects, possums, koalas, mistletoes, native cherries (*Exocarpos, Choretrum*) and dodder laurels. All of these have been blamed somewhere and sometime for causing what is commonly known as 'dieback'. (Dieback is a term that is mostly used inappropriately in Australia to refer to chronic decline, but I'll return to that.) These arbivores have struggled to exist on their diet of eucalyptus for millennia. That they stand accused of causing chronic decline is testament to ignorance amongst pathologists as well as widespread disruption of the balance of nature.

In *The New Nature: Winners and losers in wild Australia*, Tim Low wrote:

> In many parts of Australia today trees are dying and we need to know why. The immediate causes may be koalas, possums, insects,

---

VIII McIntosh described them as wet forests: see Chapter 12 for a definition of moist and wet forests.

mistletoes or diseases, but the story is always more complicated than that. A forest is a complex web of checks and balances. … That's a key point to keep in mind as we grapple with all the animals, plants and diseases that are killing our trees.[21]

In fact, these biota are taking advantage of the temporary bounty provided by dying trees. Both the trees and the arbivores are losers in the long run. This is an imbalance of nature rather than a "New Nature". Alleged overbrowsing by koalas at Cape Otway is a good example.[22] Both manna gums (*E. viminalis* ssp. *cygnetensis*) and koalas are native to the area, but the gums will soon be mostly gone along with the koalas (Fig. 5). Lack of burning (together with pasture improvement in some stands) made the trees sick. Their crowns started to decline and produce epicormic foliage. This is much more palatable and nutritious than ordinary foliage so koalas prospered. Increasing defoliation by koalas pushed the trees down the spiral of chronic decline. Now there are too many koalas and not enough leaves.

Fig. 5 Starving koala in long-unburnt, chronically declining manna gum woodland, Cape Otway. (Photo by author)

Conservation bureaucrats asserted that koalas became pestilent on Kangaroo Island because they were exotic to the island.[23] This is incorrect because eucalypts on the island have similar defense mechanisms against foliovores as do eucalypts elsewhere in Australia, Papua and Southeast Asia. Koalas were uncommon throughout their range under Aboriginal management because trees were generally healthy, and they are now overabundant in most areas containing their preferred food trees because these trees are chronically declining.

During the late 1960s there was extensive grazing and burning in eucalypt forests around Urbenville in northeastern New South Wales. Chronic decline was virtually unknown and koalas were uncommon and solitary. In 1966 J.H. Calaby famously identified a rich diversity of macropods in the upper Richmond and Clarence Valleys. Ten species of macropods occurred at Wallaby Creek, where grazing and burning maintained the natural mosaic of rainforest, savanna and grassy forest.[24] I transferred to Urbenville Forestry District in 1981 when burning was the main priority of all field staff while-ever fuel and weather conditions were suitable. There was virtually no external regulation or paperwork.

The rise of greens quickly changed the situation. I vividly recall a warm spring day, a year or two later, when conditions suitable for burning had returned. I had lit along the top of a long ridge, watched the fire trickling downslope on both sides, and gone back to the office to report my good work to the boss. He informed me that I would have to "take the gang out first thing in the morning and rake around the fire". I thought he was joking until he showed me an edict newly arrived from head office. Burning was permitted only in areas completely surrounded by rainforest, running streams, green pastures and/or roads and only if the area could be completely burnt within a single working day. As it happened, the fire had trickled down to the creek flats and gone out as the sun got lower and the humidity got higher.

Burning was easy and safe in forest and beef cattle country in northern New South Wales during the early eighties. But after the 1973 Aquarius Festival established Nimbin as the 'Hippy Capital of Australia' things were starting to take a turn for the worse. 'Ferals' and 'hobby farmers' had started buying up the 'back blocks' adjoining State forests, and fear of litigation was such that the Forestry Commission was unwilling to risk any escapes from burning in the forest. One of the 'ferals' was well known green activist Dailan Pugh, son of famous artist Clifton Pugh who, in 1989, illustrated *A Kingdom Lost: A story of the devastation of our wilderness*, which "tells the story of an Australian rainforest perilously close to extinction as it succumbs to woodchipping. ... One can almost hear the shriek of pain as man and his machines approach to begin their assault on the wilderness".[25]

Of course, Australian rainforests have never been logged for pulpwood because fibre from small, defective or misshapen eucalypt logs is far superior and cheaper to obtain as a byproduct of sawlog production. Eucalypt fibre is used to make the high quality paper that is needed for glossy productions such as *A Kingdom Lost*. In any case, Dailan bought a rainforest block that had been selectively logged for valuable timber and set up his little paradise in the 'wilderness' near Urbenville. He now rails against the Forestry Corporation, accusing it of causing the problems that he and his fellow green ideologues created.[26]

When I left Urbenville in 1985, there were a few small patches of declining forest, no more than a hectare or so in size. During five years when I spent most working hours in the bush, I had seen only one koala and had heard of two others. Within a decade, there were at least fifteen known patches of declining forest averaging nearly forty hectares in size,[27] and koalas had become the most common arboreal mammal in the region, occurring at one out of every two survey sites.[28] Incredibly, burning was further restricted from that time, on environmental grounds, at the recommendation of Forestry Commission botanist Doug Binns. Within the next decade, the area of declining forest had increased to twenty thousand hectares and koalas had become superabundant.[29]

Doug Binns, like his botanical peers, is skilled in identifying plants by close inspection of their foliage and reproductive structures. Botanists often seem to focus their attention on shrubs, perhaps because they have easily accessible and visible leaves, flowers and fruits compared to herbs and trees with less obvious or less accessible structures. During Oxley's exploration of the Western Slopes and Plains, renowned Australian botanist, Alan Cunningham, described a disproportionate frequency of shrubs compared with vegetation descriptions by several explorers in the region.[30]

Botanists seem generally averse to burning, I think because it scorches or consumes leaves and reproductive structures of shrubs, making their job harder. They also seem largely unaware of the condition of tree canopies, presumably because their focus is usually at a horizontal or lower angle. Doug and I returned to Urbenville and resurveyed some formerly healthy, grassy forests after ten years without burning. We found woody thickening, loss of biodiversity and declining canopies.[31] After this, Doug seemed to gradually change his attitude towards burning, and to start thinking more about firestick ecology. Unfortunately this happened at the wrong end of his career path and had little consequence for management of State forests in New South Wales.

I now live at Eden, where koalas have always been scarce because the forests are dominated by trees with very hard and dry leaves of low nutritional value. The carrying capacity of these forests for foliovores is consequently very low.[32] Green organisations such as Australian Koala Foundation portray naturally low numbers of koalas as an indication of environmental crisis, when it is in fact, unnaturally high numbers of this iconic animal that indicate a serious environmental problem. High populations of koalas are a sign of sick trees.[33] In the Bega Valley, European settlers sowing pastures under red gum woodlands initiated chronic tree decline and an irruption of koalas. This supported a thriving fur industry in the late-nineteenth century. As the red gums declined further, koalas suffered epidemic disease. Old trees as well as koalas in the valley were finally finished off by the Federation Drought at the turn of the nineteenth century.[34] Similar irruptions and declines occurred across much of Queensland in the early-twentieth century.[35]

By radiotracking some koalas in the forests around Eden,[36] we found that they had home ranges averaging about 170 hectares – two orders of magnitude larger than those of the superabundant koalas that have typically been studied. One mature female made a three kilometre round trip into an adjoining male's territory over two nights during the breeding season. It now makes sense to me that males bellow so loudly as to be heard several kilometres away at night time. This would be unnecessary and disruptive amongst koalas squeezed into home ranges of one or two hectares but is very useful to koalas in natural, low-density populations.

Three of our radiocollared koalas perished. A young one was speared by a falling branch, an old koala with worn out teeth died of malnutrition, and a juvenile was taken by a powerful owl – ironically another supposedly threatened species. Greens used these incidents to lobby environmental bureaucrats to stop our radiotracking, ostensibly for conservation and animal welfare reasons. Their arguments against forest management rely heavily on ignorance. They don't wish the public to know that natural, low-density populations of koalas can continue to coexist with human economy as they did for forty thousand years. In fact greens and some environmental bureaucrats have abused the Precautionary Principle to make a virtue of ignorance. This principle is: "where there are threats of serious or irreversible environmental damage, lack of full scientific certainty should not be used as a reason for postponing measures to prevent environmental degradation".[37] Naturists and environmental bureaucrats have twisted it around to stop any action whatsoever to deal with problems on the basis that there can never be complete knowledge. If one doesn't know what to do, one need not spend money on management, instead the money can be used to fund self-perpetuating research of narrow specialties.

Attitudes towards phytophthora in Victoria are analogous to those around 'koala overbrowsing' on Kangaroo Island. Plant pathologists consider it to be exotic. However, an early study by Pratt and Heather of phytophthora along the east coast of Australia from Hobart to Cairns, concluded that it was not a recent introduction because it was ubiquitous, occurring in remote, undisturbed locations.[38] The distribution of resistant plants reflected the suitability of habitats for the fungus, indicating that a balance had evolved over geological time. Furthermore disease was usually associated with post-European disturbance. Pratt and Heather were fairly confident that phytophthora was a native species at least as far south as the border of New South Wales and Victoria because it apparently caused little trouble in NSW and a lot of trouble in Victoria. Nothing much has changed over the years to alter this conclusion, but a review in 1987 claimed that "The major advances since this subject was reviewed in 1972 include the general acceptance that P. cinnamomi has been introduced to Australasia".[39] Another review in 2008 stated: "Although there has been some controversy concerning the origin of P. cinnamomi in Australia (e.g., Newhook and Podger 1972; Pratt and Heather 1973; Weste 1974), it is generally believed to be a recent arrival".[40]

Following disease epidemics in Victoria, the fungus was experimentally inoculated into supposedly healthy forests and found to cause disease.[41] But the epidemics were occurring in forests starved of mild fire, where soil conditions had deteriorated for the hosts and improved for the fungus. In other words, suitable habitat for susceptible plants was shrinking whilst the niche of the fungus was expanding. The experimental inoculations were made in very small plots (1 m²) in close proximity (45 m – 65 m) to plots containing dying vegetation and phytophthora. They simply accelerated the spread of the fungus in its expanding niche. Megafires also favoured the fungus because they interrupted transpiration by trees and bushes, raising watertables on poorly drained sites and facilitating its growth and dispersal. This was a 'triple whammy' because rising water tables suffocated the roots of trees that had been defoliated by fire.[42] Poorly drained sites with heavy mortality and phytophthora have come to be known as "graveyards" in the media[43] and in pathological literature.

Pathologists consider that:

> One of the intriguing dieback puzzles is the absence of disease from wet sclerophyll forests, which are often dominated by susceptible species and occur near sources of infection, e.g., the Victorian forests of mountain ash (*E.regnans*). Wet sclerophyll or tall closed forests grow on suppressive soils in mountainous regions where rainfall is high and reliable.[44]

Disease doesn't occur in **true** wet sclerophyll forests (see Chapter 12) because their fire regimes and soils haven't changed since European settlement.[45] The trees aren't chronically declining and their roots aren't susceptible to the fungus.

Another puzzle intriguing pathologists such as David Cahill and colleagues, is that Phytophthora has only recently become a problem in New South Wales. They tell us that:

> Although *P. cinnamomi* was first associated with plant deaths in native vegetation in NSW in the late 1940s (Fraser 1956), it has, until recently, not been regarded as a significant threat to native vegetation in NSW. The perceived low threat of *P. cinnamomi* was based on a conclusion reached by Pratt and Heather (1973), since dismissed, that the pathogen was native in eastern Australia.[46]

In 1979, the Secretary of the Forestry Commission of NSW responded to a newspaper article titled *Graveyard of Our Gum Trees* which claimed that phytophthora was "killing large areas of trees in the forests of NSW, Victoria and Queensland, but especially in the great jarrah forests of Western Australia". Secretary A.R. Cocks (who coincidentally inducted me as a professional forester the year before) pointed out that:

> That general statement is not true ... NSW ... pathology staff were among the pioneers (in the 1950s) in studying this root rot disease in Australia. ... in spite of its very common occurrence in NSW forest soils we have yet to record any widespread areas of death attributable to it. ... we have stepped up monitoring operations in NSW forests but fortunately we can still regard this common fungus as benign.[47]

Pratt's and Heather's conclusion probably shouldn't have been dismissed so lightly three decades later. Cahill and colleagues moved from a vague general belief that phytophthora is exotic, to complete dismissal of the alternative view, without presenting any additional evidence. Indeed they were puzzled by some of their own data that support the conclusions of Pratt and Heather: "Oddly, *E. baxteri*, a species severely affected by *P. cinnamomi* in the Brisbane Ranges of Victoria, shows no symptoms of disease on infested sites in NSW (such as Green Cape)".[48] These observations suggest that phytophthora is indeed native to the east coast, where there has been sufficient time for tree species to develop resistance or be lost from habitat suited to the fungus. Perhaps it is a more recent introduction to the southern coast.

On the other hand, the Brisbane Ranges National Park shows all the signs of chronic decline due to long-term lack of frequent mild fire. There are dense understoreys or thick litter layers, as well as plagues of mistletoes in some areas and plagues of koalas in other areas that are apparently unaffected by phytophthora. Phytophthora appears to be just one of a suite of native arbivores that has irrupted as a result of disturbance.[49] It has likely been in eastern Australia longer than dingoes because they probably arrived only three or four thousand years ago, long after the most recent separation of Australia from New Guinea by rising seas.[50]

Cahill and colleagues argued that:

> *P. cinnamomi* has been isolated from numerous sites along the coastal fringe (O'Gara et al. 2005) but impacts on natural flora and vegetation attributable to *P. cinnamomi* are localised and few. Given that these impacts have only recently been discovered and that *P. cinnamomi* is present in many areas where diseased plants have been observed, *P. cinnamomi* may prove to be a significant threat to biodiversity in NSW. Further survey and research is required.[51]

In fact localised impacts have been studied since the 1950s,[52] and they have become more numerous and extensive in national parks on sandstone country around Sydney since broadscale fire suppression commenced in the 1970s.[53] Fire suppression and megafires since that time have created a bonanza for the fungus which is taking advantage of declining health in iconic species such as smooth-barked apples and grass trees.[54] Now that stark white skeletons of dead trees overtop dark green scrubs in highly visible locations around Sydney Harbour, park managers and pathologists have realised that there is a problem.[55] But it is **not** phytophthora. Sensible management is required to deal with it, rather than "Further survey and research". Unfortunately the problem is so advanced in many places that substantial clearing is required before it can be managed with fire.

Next to phytophthora, Mundulla Yellows (MY) was probably the 'disease' of greatest interest to forest pathologists in recent times. They spent decades from the 1970s searching for a virus or phytoplasma[IX] to blame for this type of tree decline. However, by 2007, Parsons and Uren had established that "All of the well-documented records of MY are from sites where acidic soils have become more alkaline because of the use of crushed limestone in road-making or to the use [sic] of alkaline irrigation water".[56] Three years later, prominent pathologists were still searching for biotic 'causes' of a condition that had been clearly established as lime-induced chlorosis (yellowing and dysfunction of leaves).[57]

Bell-miner-associated dieback – a brilliantly clumsy epithet – is perhaps the extreme case of the search for the holy grail of biotic disease. It is just another example of chronic decline associated with lack of burning, but in this case the pests that proliferate in sick trees are psyllids. These sap-sucking insects feed on eucalypt leaves, and bellbirds like to eat them. Psyllid plagues occurred in moist forests along New South Wales' central and north coasts in the 1940s and 1950s after decades of fire suppression. Forest entomologist K.M. Moore investigated chronic eucalypt decline in Ourimbah State Forest between 1956 and 1960.[58] (Pathologists also attributed this case of decline to phytophthora.[59]) Bellbirds were feeding on plagues of psyllids, but Moore sensibly looked into weather and soil conditions, rather than birds, in searching for the cause of the problem. He noted that: "During the early stages" of an irruption of psyllids "damage is not readily discernible from the ground. The first indication of psyllid attack may be the presence of bell-birds".

Moore noticed that there was often a sharp upper limit to the area plagued by psyllids that fell short of the: "limits of host distribution. This was sometimes denoted by the presence of a road, an abrupt rock-face, or a steep incline, beyond which the psyllids did not persist in large numbers". He failed to appreciate that such upper limits usually correspond to the lower limits of ridgetop fires. Moore found that the areas attacked by psyllids had "shallow topsoil (11" to 18") over deep, heavy clay". These were soils that had developed with frequent burning[60] and had become

IX Microscopic bacteria that feed on conductive tissues in plants.

unsuitable for fire-dependent trees after fire was suppressed. Bellbirds extended their range into the Watagan Mountains in the 1960s[61] after psyllid plagues irrupted in the mountain gullies. Broad-area burning was introduced to State forests in New South Wales about the same time in response to megafires and to chronic declines that caused irruptions of stick insects on the tablelands as well as of psyllids on the coast. Fire safety and forest health improved temporarily. Burning was reduced from the 1980s, and plagues of psyllids, stick insects and other arbivores reappeared from the early 1990s.[62]

In 1981 Richard Loyn and colleagues famously removed bellbirds from a stand of eucalypts in Victoria that was plagued by psyllids.[63] They reported that other birds flocked in to eat psyllids, and the canopy condition of the trees improved. This was the genesis of a new hypothesis that bellbirds caused eucalypt decline by 'farming' psyllids – selectively harvesting lerps (sugary exudates of the psyllids) and larger nymphs whilst excluding other predators. However Loyn failed to explain why irruptions of bellbirds lagged behind irruptions of psyllids and that birds irrupted to similar levels whether they were bellbirds alone or a number of species including bellbirds.[64] This study actually demonstrated that bellbirds can remain 'sedentary' because they are inefficient predators whereas more efficient avian predators of psyllids operate as nomadic flocks that move on to exploit other infestations after they have temporarily exhausted the resources at a particular site.

Nevertheless it became 'common knowledge' that bellbirds cause chronic eucalypt decline by farming psyllids. When New South Wales' forest entomologist, Dr Christine Stone, started investigating psyllid outbreaks in the early 1990s,[65] she discounted about a third of the farming hypothesis – the part where the birds only choose the lerps, because this had been disproven by Poiani in 1993.[66] Stone hung onto the bit where the birds keep other predators at bay: "the presence of bell miners is highly correlated with this dieback problem … This could possibly occur because of their harassment of other insectivorous birds". She further suggested that bellbirds might interfere with or eat psyllids' invertebrate predators.

Stone also wrote that:

> The increasing spread of canopy dieback may ... be linked to
> maintenance ... and/or increase in bell miner territories. ... A
> salient feature of dieback-affected areas occupied by bell miners
> is the presence of gaps in the overstorey, resulting from either
> dead standing trees or contracting mature crowns. ... We therefore
> postulate that bell miners prefer a dense understorey and a
> discontinuous sclerophyll overstorey.

Four decades after Moore had observed and reported that psyllids
built up before bellbirds arrived, and following seven years of her
own research, Stone published a model wherein bellbirds cause psyllid
plagues. She claimed that opening of the forest canopy by logging,
roading or fire allowed dense understoreys to develop; this provided
attractive nesting sites for bellbirds which more or less farmed psyllids;
and these psyllids made the trees progressively decline from the canopy
down to the roots.[67] At the same time, a repeat of Loyn's experiment by
other ecologists showed that removing bellbirds allowed other birds to
dramatically reduce psyllids, but there was no improvement in tree health.
The authors concluded that psyllids weren't the fundamental problem.[68]

Plagues of psyllids and bellbirds are a consequence rather than a cause
of canopy decline. I have observed the progress of chronic decline with
and without bellbirds throughout south-eastern Australia for a decade and
a half. It is quite obvious that canopies decline well in advance of, or even
without any colonisation by bellbirds. I discussed these observations with
my friend, colleague and former boss, John Turner, and he explained to
me the soil processes that are at the root of the problem. We published
a simple model of 'dieback' based on my observations and John's
knowledge of soils and nutrient cycling processes according to different
fire regimes. Tree roots and mycorrhizae deteriorate with changing soil
conditions in the absence of fire. This leads to canopy decline which
allows understoreys, psyllids and bellbirds to flourish. Epicormic foliage
in declining canopies is attractive and nutritious to psyllids which are
attractive and nutritious to bellbirds.[69]

Having received awards to observe chronic eucalypt decline more widely across Australia, and having examined overseas literature on chronic decline, I published a more general concept in an international journal.[70] With John Turner and other colleagues, we also analysed soils to confirm the undergound processes.[71] Meanwhile Stone and her colleagues spent lots of time and resources confirming an association between bellbirds and chronically declining forest on New South Wales' central coast. They apparently thought that they were demonstrating cause. In 2008, 15 years of research culminated in an article in *Australian Forestry*.[72] The key point in the *Summary* was: "In this study site, the presence of bell miners was significantly associated with unhealthy eucalypt crowns. This supports the proposition that a dieback syndrome known as bell-miner-associated dieback (BMAD) exists in central coastal forests of NSW". I weep for logic, science, the English Language, and most of all, for our mismanaged, chronically declining forests.

To make matters worse, the underlying concept that co-evolved arbivores can destroy their food trees remains current in pathology and academia. For example, in 2013 the New South Wales Government allocated nearly half a million dollars to research the genetics and population dynamics of eucalypts, psyllids and predators of psyllids in Cumberland Plain Woodlands. This research is intended "to develop novel management strategies for psyllid outbreaks on eucalypts".[73] There will be no return on this investment – the money should be spent on burning – a tried and tested management strategy.

In 2010, an invited feature article published by the Ecological Society of America,[74] and synthesising a "global assessment" made it clear that "N accumulation is the main driver of changes to species composition across the whole range of different ecosystem types by driving the competitive interactions that lead to composition change and/or making conditions unfavorable for some species." The article succinctly summarised the process underlying chronic decline as follows:

> resistance to plant pathogens and insect pests can be lowered because of lower vitality of the individuals as a consequence of

N-deposition impacts, whereas increased N contents of plants can also result in increased herbivory. Furthermore, N-related changes in plant physiology, biomass allocation (root/shoot ratios), and mycorrhizal infection can also influence the susceptibility of plant species to drought or frost.

Contrast this logical science against the following statement by some members of The American Phytopathological Society in 2011:

> forest pathologists use "decline" to describe forest tree diseases of complex etiology. We contend that this distinction from abiotic or biotic diseases is completely arbitrary, has caused undue confusion, and provides no practical insights for forest managers. ... We propose that forest pathologists discontinue the use of "decline" as a distinct category of disease. Furthermore, we suggest that new diseases should be named based on the affected host, characteristic symptom, and, once known, major determinant. We believe that clearer communication in describing complex diseases is a prerequisite to finding effective management options.[75]

I have no problems with the protocol suggested by Mike Ostry and colleagues for naming diseases, but they aren't really diseases and they aren't really complex. Following their suggestion, most supposedly complex diseases in Australian and North American forests would be called "tree decline from lack of burning" and those in northern Europe would mostly be "tree decline from atmospheric pollution".

Plant pathologists have long paid lip service to "The Disease Triangle". This is an equilateral triangle illustrating that expression of disease is equally dependent on the coincidence of three factors – a susceptible host, a virulent pathogen and a suitable environment. However the concept doesn't recognise that there are often multiple biotic factors. Nor can it accommodate an 'abiotic disease' where there are only two factors – 'host' and environment. In fact, there is really only one factor that is important in 'disease' and that is disturbance. Disease is just another word for an imbalance of nature, where a fundamental change in the environment causes immediate harm to some organisms and temporarily benefits others until such time as they once again hit the wall of resource limitation. Forest pathologists will not find any solutions to irruptions of pests, parasites and diseases until they recognise this fact.

T.C.R. (Tom) White is a pathologist who understands that co-evolved native species cannot cause decline of healthy trees. In *The Inadequate Environment: Nitrogen and the abundance of animals*, he took great pains to establish "the universal hunger for nitrogen as the misery that drives the ecology of all organisms".[76] Unfortunately he does not appreciate the critical importance of fire in maintaining a balance of nature by preventing the accumulation of nitrogen in soil. Excess nitrogen temporarily provides unlimited resources for 'pestilent' species. White proposed weather as the primary cause of "diebacks and declines". He claimed that:

> declines and diebacks are but one extreme of a continuum of response of trees to physiological stress; at the other extreme are small, short-lived increases of predators [arbivores] on one or a few trees. Outbreaks of insects and fungi of varying duration and severity fall between these two extremes.[77]

Actually, dieback can be a healthy response to natural stress such as extreme weather, whilst decline is an unhealthy condition brought on by disturbance. To understand chronic decline, it must be distinguished from natural dieback, and this can be difficult in its early stages. However canopy condition can be monitored until after a natural stress is alleviated. Epicormic shoots on healthy trees grow into new, normally developed crowns when natural stress ends. Declining trees continue to shoot and reshoot epicormically.[78] White suggested that 'New England Dieback', an example of rural tree decline, was a result of natural stress to trees from repeated waterlogging and drought.[79] In fact, it is a result of unnatural accumulation of nitrogen and associated soil changes, usually as a consequence of pasture improvement. The quantity of foliage on the declining trees varies with seasonal conditions, but normal crown development does not resume in favourable seasons.

Chronic decline has often been confused with drought stress in Australia, but the two processes produce different patterns in the landscape. Droughts cause dieback in dry sclerophyll forests on shallow soils and exposed aspects, whilst woodlands in lower, sheltered positions are unaffected. During good seasons mismanaged woodlands continue to decline whilst adjoining dry sclerophyll forests remain healthy. Drought sometimes causes dieback in Australia, but it has been used as a scapegoat to explain chronic declines.[80] Drought stress affects trees, saplings and understoreys. It is most pronounced in dense stands. Chronic decline usually doesn't affect saplings, and understoreys mostly thrive. Severity of decline is independent of tree density.[81] Drought stress is more severe in subdominant trees than in dominant trees, whilst decline affects dominants as much as any other trees. Thinning alleviates drought stress, but can exacerbate decline.[82] Perhaps the most compelling evidence against climatic stress as a primary cause of decline is in obvious contrasts between the health of similar stands on adjoining properties. Frequently burnt or grazed or slashed stands are healthy whereas ungrazed or infrequently burnt stands, and trees in improved pastures are unhealthy.[83]

Paul D. Manion in North America came up with a reasonable concept of decline which he rather clumsily named "The Decline Disease Spiral".[84] Even clumsier was his classification of predisposing, inciting and contributing factors, wherein natural and unnatural factors were mixed, as were biotic and abiotic factors. Little wonder that some of his colleagues, led by Allan N.D. Auclair, described the concept as a "perceptional miasma".[85] Unfortunately Auclair threw the baby out with the bathwater, claiming that climatic cycles drive 'dieback', even though his own data showed an ongoing upward trend of canopy decline superimposed on a 22 year climatic cycle.[86] Manion's inability to explain this same trend led him to the rather odd suggestion that forests need a "healthy amount of disease".[87]

Manion tried to turn Darwin's *Theory of Evolution* upside down. He claimed that "Periodic inciting events and subsequent decline in canopy dominant trees provides an opportunity for the less competitive individuals to contribute to the gene pool for future generations. Decline diseases therefore contribute to the stability of the forest system". But dominant trees in healthy stands are resilient to extreme events, whilst dominant and subdominant trees in declining stands are equally vulnerable. Manion's "decline disease stabilising selection concept" is as logically clumsy as it sounds. It is an attempt to explain unnatural decline caused by disturbance as if it were a natural process. He appears to have confused shade-tolerance of trees with tolerance to "stress": "One concept categorises life strategies for species of plants into competitive dominants and stress-tolerant dominants. Different individuals within a species can also be characterised as having competitive dominant or stress-tolerant dominant tendencies".

In fact, naturally subdominant, shade-tolerant species proliferate when intolerant dominant species decline as a result of disturbance such as fire suppression. When mature trees decline in uneven-aged stands, younger trees of the same species can also gain a competitive advantage. These processes changed mixed open woodlands into cypress thickets such as the Pilliga Scrub in New South Wales, and also turned open river red gum woodlands on the Murray River into scrubs. But it was thinning by foresters and graziers, not tree decline, which turned the scrubs into healthy forests.[88]

To clear the perceptional miasma and confront the reality of chronic decline, the factors involved can be consistently classified as follows:

- **Predisposing factors are genetics and site**. For example, soils developed under frequent Aboriginal burning favour trees adapted to low nutrient, poorly drained sites. Excluding fire from these sites changes the soils so that they are no longer suited to the trees.

- **Inciting factors are always physical and/or chemical stresses** caused by human disturbance such as fire suppression, fertilisation or pollution. Natural weather extremes in healthy ecosystems lead to temporary dieback of trees and temporary flushes of pests and diseases followed by recovery.

- **Contributing factors are biota** including arbivores that respond to improved food quality in sick trees, and competitors that take advantage of increased light, reduced competition and increased resources under declining canopies. They cannot flourish spontaneously.

These differences between natural stress and chronic decline were evident from an unintentional experiment in a forest near Urbenville.[89] An open, grassy, regrowth eucalypt stand straddled a boundary fence. A private owner grazed cattle on one side whilst grazing and burning were excluded from the State forest side in the early 1980s to 'protect' the regrowth trees. The stand was healthy on both sides of the fence when I left the area in 1985. When I returned fifteen years later, the trees on the privately owned side were still healthy whilst the trees in the State forest were dead or dying. The remaining live trees were plagued by psyllids and a dense understorey of native and exotic shrubs had developed. Bellbirds were thriving on the psyllids. A drought was developing, and as it intensified, the trees in State forest lost most of their leaves whilst the canopies of the privately owned trees grew thin. Psyllids and bellbirds abandoned the dying trees and moved into the drought stressed trees on private property. After drought breaking rains, the trees on the private land recovered and surviving trees in the state forest regained sparse epicormic foliage on their branches. Psyllids and bellbirds retreated into the State forest.

There were no obvious arbivorous contributors to 'high altitude dieback' of alpine ash in Tasmania when foresters first noticed it. Dense green undergrowth suggested that drought wasn't the problem, but the understorey might have been contributing to 'dieback'. Bob Ellis and colleagues felled the understoreys in a few stands but they didn't improve. On the other hand, felling and burning the understorey restored the health and vigour of treated stands. Ellis worked out that lack of burning caused problems in the soil that made the trees sick. Later on, he found out that eucalypt roots and mycorrhizae[X] were unhealthy and deformed in soils under long-unburnt stands.[90]

X Mycorrhizae are fine root 'hairs' consisting of fungal tissue in symbiosis with tree roots.

Recently the Bushfire Cooperative Research Centre (CRC) built upon this seminal work and extended it to investigate tuart decline in Western Australia. In both cases, lack of fire led to accumulation of soil nitrogen in one form or another. In the case of alpine ash, the soil derived from granodiorite became more acid whereas pH was unaffected in the limestone derived soil under tuart. Changes in soils caused nutrient imbalances and declining health of trees on long-unburnt sites.[91] Unfortunately the CRC wasted a lot of resources on very specialised research at the same sites.

Dugald Close and colleagues had formulated an hypothesis in 2009 that midstorey development in the absence of fire deprives trees of water and nutrients causing chronic decline.[92] Ellis's work had previously shown that this was not the case. I took Close and his colleague Neil Davidson on a short tour of the bush in the south-east mainland and pointed out that eucalypt canopies deteriorate first, and then understoreys develop. They were not impressed by my observations and examples. Instead they set up an expensive project which measured carbon isotope ratios in the leaves of healthy and declining tuart trees to compare the efficiency with which they used the water from the soil. Healthy and declining trees used water differently[93] but this didn't prove that the understorey was the problem. When the understorey was cleared it made no difference.[94] The experiments were a waste of time and money because Ellis had already tested and rejected the hypothesis.

Taxpayers, through the CRC, also paid for specialised research of changes in mycorrhizal communities in the absence of fire. However, it had been recognised forty years previously that increasing nitrogen in soils can cause deterioration of roots and mycorrhizae,[95] and Ellis and Pennington had shown in the early nineties that this was driving decline of long-unburnt alpine ash stands.[96] Bryony Horton and colleagues reported their 'new' research in 2013: "This study of eucalypt forest decline is the first to investigate the link between ECM fungal community attributes and health of the forest trees." This claim contradicted the following statements in the introduction to their paper:

> The percentage of mycorrhizal roots on *E. delegatensis* seedlings grown in soils collected from declining and healthy forests differed

Rarity, decline and pestilence   139

in abundance and form (Ellis and Pennington, 1992). Seedlings grown in a mix of soil collected from healthy and declining forest formed prolific mycorrhizas with fungi of the Basidiomycota, whereas seedlings grown purely in soil from declining forests formed less abundant mycorrhizas with the Ascomycota (Ellis and Pennington, 1992).[97]

Horton is one of many narrowly specialised academics in Australia who, in my opinion, seem to strive for ongoing specialist research rather than outcomes that deliver improved environmental management. Although she and her colleagues concluded that "Forest management must consider the role of disturbance [burning] in maintaining suitable soil conditions for symbiotic fungi whitch are important for maintaining healthy eucalypt forest and restoring declining forest ecosystems", they failed to deliver this outcome.

Horton was appointed by New South Wales National Parks and Wildlife Service in 2012 to lead a project called *Mitigating the effects of Bell Miner Associated Dieback* in World Heritage Areas. This project was funded by a Commonwealth Government grant scheme known as *Caring For Our Country*. I spent a long day showing her all aspects of the history and current extent of chronic eucalypt decline in World Heritage Areas around Toonumbar and Urbenville in northern New South Wales. This included Wallaby Creek where the diversity of vegetation and macropods was rapidly succumbing to scrub invasion. I was hoping to impress upon her the urgent need to reintroduce frequent mild fire in the landscape to restore the world heritage values. Her response at the end of the day was to repeat the typical mantra of academics. She said that "we need to do some more research".

It was quite clear from the initial output of the project – a two page document attempting to resuscitate disproven hypotheses and referring to trials of "bell miner habitat modification"[98] – that none of the resources of this project would actually go to 'caring for country'. This, it seems, was confirmed in 2015 with the publication of more research from this project. Unsurprisingly, Martin Steinbauer and colleagues found that there was more nitrogen in soft young leaves than tough old leaves, and that there were correlations amongst foliar nitrogen, psyllids, animals that eat psyllids and animals that eat animals that eat psyllids.[99] Also unsurprisingly, they called for further research.

Walter Starck said in relation to the climate debate:

> When large amounts of money are made available to study a problem the one thing least likely to be discovered is that there actually isn't one. Almost certainly it will be found that the problem is complex and more research is urgently needed.[100]

It seems just as unlikely that ecological research will report a simple solution to a real problem.

General ignorance of the basic process of chronic eucalypt decline with lack of fire has led to even greater waste of resources in other fields.

I visited Hawkesbury Institute for the Environment (HIE) in 2013, after they had spent twenty million dollars setting up a "Free Air CO_2 Enrichment (FACE) experiment, located in an existing, natural evergreen woodland with low levels of soil mineral nutrients".[101] Unfortunately the 'natural woodland' is really an unnatural forest that has never seen fire. The tree density, at 650 to 900 stems per hectare, is one or two orders of magnitude higher than the density of natural red gum woodlands. The trees are only 60 or 80 years old and they are chronically declining. This is clearly apparent in glossy brochures and web pages produced by HIE[102] showing sparse canopies of epicormic foliage in the trees. The level of nitrogen in the soil is undoubtedly elevated as a result of altered nutrient cycling, and the grassy sward is unnaturally dense and homogenous. This elaborate and costly experiment will produce data that have no relevance to the growth of natural woodlands.

The Bushfire Cooperative Research Centre has been disbanded, and pathologists at the *Centre of Excellence for Climate Change Woodland & Forest Health* in Western Australia now lead specialist research into chronic eucalypt decline. All the funds that continue to be 'splurged' on this issue merely inhibit action on the obvious solution because specialist researchers are looking for funding rather than practical management outcomes. Meanwhile, misplaced concerns about prescribed burning are inhibiting action to save species and whole ecosystems such as red gum woodlands that are being pushed towards extinction by disturbance to the balance of nature. They are being choked by native shrubs and exotic plants whilst changing soil conditions and declining health of trees promote pests, parasites and diseases.

A local population of grasstrees in Royal National Park was recently extinguished after a severe fire and outbreak of phytophthora following two decades of fire suppression. Grasstrees have been declining around Sydney since monitoring commenced a quarter of a century ago.[103] Broad-headed snakes are extinct in Ku-ring-gai National Park, and declining in Morton National Park because shrubs are obliterating their basking areas.[104] The eastern brown treecreeper disappeared along with its open grassy habitat, when two State forests in northwestern NSW were converted to national parks[105] and 'protected' from grazing and burning. Hastings River mice survive in grazed and burnt forest, but have been lost from 'protected' areas in national parks in northeastern New South Wales[106] where their grassy habitat has been choked out by scrub and overrun by bush rats and antechinus. Incredibly, the environmental bureaucracy has banned grazing and burning from the habitats of these mice in State forests.[107] It consistently regulates to 'protect' the saplings, shrubs, fallen timber and parasitic plants that are taking over the landscape at the expense of ancient trees and rare plants and animals.

The most dramatic faunal decline in Australia during my lifetime has been that of the formerly common and widespread Tasmanian devil. When cyclone Tracy levelled Darwin in the Christmas of 1974, I was enjoying vacation employment as a chainman on forest assessment far away in eastern Tasmania. My job was to drag a 100 metre steel survey band – known as a chain – on a compass bearing through the bush and establish a systematic grid of sample plots where the assessor at the other end of the chain would record measurements to describe the forest. Often, when I left the transport vehicle in the morning, I carried my tools, food and drink for the day. It was mostly easy walking through open forest and I rarely needed to use my brush hook to clear a line of sight or passage. Devils are mostly nocturnally active, but I saw one in the bush and several along the roads at night. Devils killed by vehicles on the east coast highway were a very common sight. The estimated population[108] at that time was about 100,000 devils.

Thirty years later, the generous support of the Joseph William Gottstein Memorial Trust allowed me to return to the east coast of Tasmania to examine the problem of chronic eucalypt decline. Tasmanian forestry pathologist, Tim Wardlaw had kindly agreed to show me examples of 'gully dieback' and 'east coast dieback'. I was not at all surprised to find that these supposed 'diseases of complex etiology' were obvious manifestations of chronic decline consequent to lack of burning.[109] (Pathologists attributed them respectively to drought stress compounded by armillaria; and drought stress followed by waterlogging stress and exacerbated by phytophthora.) However, I was horrified to find that the open forests of my youth were no more. I couldn't imagine anyone attempting to cut a line on a compass bearing through the new scrub and assess the dead and dying trees for timber. Nor could I understand how the extent of scrub invasion and tree decline could go unnoticed by the authorities and the general public.

Later, I wasn't the slightest bit surprised to learn of the population crash of Tasmanian devils and the severe decline in eastern quolls, because their habitat was gone with the open forests. The estimated population in eastern Tasmania is now about ten percent[110] of what it was when the east-coast forests were mostly grazed and burnt. In my opinion, the facial tumour disease in devils, the various 'new' phytophthoras and other fungi that are occupying the plant pathologists of the Centre of Excellence for Forest Health in Western Australia[111] and the 'new' psyllid attacking eucalypts on the Cumberland Plain[112] are distractions from the fundamental issue. Eucalypts and devils are diseased and declining because their habitats have disappeared – not cleared, but changed beyond recognition since mild fire has been excluded from the landscape.

There is every reason to suspect that the chytrid disease plaguing frogs in Australia, including critically endangered corroboree frogs, 'rows in the same boat'. It's spread by common and widespread frogs that are immune to this disease[113] and are presumably favoured along with the fungus by lack of burning and woody thickening. Corroboree frogs in the Alps have been declining since about 1980, and ecologists put the decline down to habitat damage by grazing; increasing wildfires as a result of climate change; and disease.[114] The last, limited grazing (for drought relief) was in 1973,[115] however wildfires certainly increased after prescribed burning was deliberately reduced by green bureaucrats from the mid-seventies.[116] Most of the corroboree frog's boggy habitat was burnt by the 2003 Alpine megafires, putting it "in real danger of extinction".[117]

Without disturbance, co-evolved diseases and hosts remain in balance. Exotic chytrid disease has apparently extinguished many species of frogs because it can survive in other, more resistant species. But devils' cancer lives in no other species. Diseases restricted to a single host usually reach a balance as host populations are thinned out, transmission is reduced, and resistant genotypes survive. According to Hamish McCallum, devils' cancer is virtually unique in so far as it is a sexually transmitted disease where transmission is independent of host density, it cannot develop reduced virulence because it is clonal, devils' immune system doesn't recognise it, and this is unlikely to change because of devils' lack of genetic diversity.[118]

McCallum considered that the possible extinction of the devil illustrates some important principles about disease and conservation biology. I find it hard to reconcile with the theory of evolution. It's hard to imagine how an organism could be fit enough to reduce its host population by 90% in a short space of time but not fit enough to achieve its own long term survival by reaching a balance with its host. Time will tell. In the meantime it illustrates a fundamental principle of land management for conservation, and the disastrous consequences of true disturbance.

Tasmanian Devils prefer open forests and woodlands mixed with grasslands[119] rather than dense and wet forests, and they kill wombats, macropods and possums for food.[120] They take possums mostly on the ground, whereas spotted tailed quolls are more successful possum hunters because they catch them up in trees.[121] So, my hypothesis is – less burning, more trees and shrubs, less grass and browse for devil prey, less possums on the ground – more work for less food – chronic stress – disease – decline.

But I was shocked when I read in a scientific paper that devils spend little time foraging in dry compared to wet eucalypt forest.[122] This contradicted my hypothesis and previous articles that I had read about devils. When I turned back to the description of the study area, 'the penny dropped'. I found that dry eucalypt forest has a "closed, dense shrub layer from 0 to 2 m" whereas wet eucalypt forest has a "closed shrub layer at 2-6 m and a moss ground layer". It is clear that the new scrubs replacing the open forests are unsuitable habitat for devils and their ground-dwelling prey. The devils are struggling on meager resources in unsuitable habitat. The only reason I can see why devils would spend a lot of time in 'wet' forest is because they can walk through it more easily than dry scrub. The 2013 megafires on the east coast of Tasmania were an inevitable consequence of this scrub development and more will follow.

A surprising aspect of the devils' crisis is the relative lack of publicity on the mainland compared to 'environmental disasters' attributed to man-made climate change and regulation of the rivers in the Murray -Darling Basin. It seems that greens have so far been unable to dream up a link between devils' disease and capitalism. Otherwise the devil would be a pinup species for wilderness enthusiasts. As a top level predator it would fit the bill as an 'indicator species' of the alleged ecological evils of human economy.

The successive declines of mesofauna in New South Wales' Western Division, in Australia's Western Deserts and in Tasmania have one common element – disruption of sustaining fire. But academics and conservation bureaucrats continue to campaign against burning, and to promote saplings, shrubs, and litter as 'habitat features' that have supposedly been reduced by modern human economy. These 'features' are taking over the landscape, and obliterating the real features that sustained biodiversity.

Dr. David Watson's fascination with mistletoes provides an extreme example of pestilence promoted as biodiversity. Mistletoes were uncommon under Aboriginal management. Amongst thousands of plants from more than a hundred different species noted by Oxley, Sturt and Mitchell in southwestern New South Wales, there was but one mention of mistletoe.[123] Aboriginal burning controlled mistletoes by maintaining healthy trees which resisted their establishment. Seedlings could only establish on a limited number of subdominant or otherwise stressed trees that weren't able to resist penetration by the mistletoe's haustorium or 'root'. If they establish on healthy trees suffering drought stress, the mistletoes usually die because they are profligate users of water,[124] whereas the trees firstly die back and then recover after drought-break.

Dr. Watson counted birds in a stand where mistletoes were removed and another stand where they weren't. He found more species (61) in the stand with mistletoes compared to the one without (52).[125] He went on to produce a great volume of literature extolling the virtues of mistletoes as enhancers of biodiversity, nutrient cycling and heterogeneity of habitat. But historical records show that small birds and mistletoes were both uncommon because Aboriginal burning maintained healthy trees with few bugs for birds to eat, and also destroyed nests. Watkin Tench wrote that Aborigines ate eggs whenever they could get them.[126] Charles Darwin would be astounded to hear that anyone had counted 52 different birds in one place on the Tablelands of New South Wales. During his visit to the Central Tablelands, Darwin remarked "In these woods there are not many birds".[127] But, because unnaturally high densities of mistletoes temporarily support high numbers of birds, Watson described them as environmental keystones (not much of them, but critically important). He eventually progressed from architectural analogy to Greek Mythology, describing mistletoes as Dryads – deities that depend on their special trees and create sacred groves around themselves.[128]

The mythological flavor was quite appropriate. Watson reported that he was working in "open woodlands" with more than 800 trees per hectare – actually an order of magnitude denser than natural woodlands. There were nearly 100 mistletoes per hectare[129] – probably two orders of magnitude more than in natural woodlands, and symptomatic of chronic tree decline. One wonders how the various species of birds survived for forty thousand years without plagues of mistletoes and other arbivores to sustain them. A more appropriate architectural analogy would be that the mistletoes are a network of cracks in the plaster indicating problems in the foundations. Mythologically they could be seen as a Hydra guarding the entrance to the underworld and sprouting two heads for each one cut off, thus reflecting pathologists' obsessions with complexity, and inability to come up with solutions dealing with the simple root causes of chronic tree decline. Keystones cannot remain floating in the air when foundations crumble.

Arbivores, saplings, shrubs, litter, fallen timber, scrub, bush rats, cockroaches, common birds, parasites and diseases are proliferating at the expense of healthy ancient trees, grasses, herbs, bare ground and the animals that depend on them. European settlers disturbed the balance of nature when they stole the firestick.

Endnotes

1   Darwin 1860.
2   James and McDougall 2007.
3   Cook *et al.* 1992.
4   James and McDougall 2007.
5   Environment Australia 2008.
6   Pyne 1991.
7   James and McDougall 2007.
8   Bombala Times 2011.
9   Jurskis 2005.
10 Jurskis and Walmsley 2011, Stanton *et al.* 2014a.
11 Montreal Process Implementation Group for Australia and National Forest Inventory Steering Committee 2013.

12 Howitt 1891.

13 Ellis *et al.* 1980, Ellis and Pennington 1992.

14 Landsberg *et al.* 1990.

15 Jurskis 2005.

16 Wolfe 2009.

17 Jurskis 2005.

18 Gibbons *et al.* 2008.

19 Jurskis 2005, Turner *et al.* 2008, Jurskis *et al.* 2011.

20 McIntosh *et al.* 2005.

21 Low 2003.

22 ABC News 2012.

23 Natural Resource Management Ministerial Council 2009.

24 Calaby 1966.

25 Pugh and Blashki 1989.

26 Pugh 2014.

27 Stone *et al.* 1996.

28 State Forests of NSW 1995, Table 11-3.

29 Jurskis *et al.* 2003, Jurskis 2005.

30 Croft *et al.* 1997.

31 Jurskis *et al.* 2003.

32 Braithwaite *et al.* 1984.

33 Gordon and Hrdina 2005, Jurskis 2005.

34 Lunney and Leary 1988.

35 Gordon and Hrdina 2005.

36 Jurskis and Potter 1997.

37 Commonwealth of Australia 1995.

38 Pratt and Heather 1973.

39 Weste and Marks 1987.

40 Cahill *et al.* 2008.

41 Weste 1974.

42 Fagg *et al.* 1986.

43 Cocks 1979.

44 Cahill 2008.

148   Firestick Ecology: Fairdinkum science in plain English

45 Turner 1984, Jurskis 2005.

46 op. cit.

47 Cocks 1979.

48 op. cit.

49 Jurskis 2004.

50 Barlow 1981, Newsome 1983.

51 op. cit.

52 Cocks 1979.

53 Conroy 1996.

54 Jurskis and Underwood 2013.

55 Sydney Harbour Federation Trust 2004.

56 Parsons and Uren 2007.

57 Parsons and Uren 2011.

58 Moore 1961.

59 Pratt and Heather 1973.

60 McIntosh *et al.* 2005.

61 Higgins *et al.* 2001, p. 611.

62 Jurskis *et al.* 2003; Jurskis 2005, 2011a.

63 Loyn *et al.* 1981.

64 Ibid., Fig. 1.

65 Stone *et al.* 1995.

66 Poiani 1993.

67 Stone 1999.

68 Clarke and Schedvin 1999.

69 Jurskis and Turner 2002.

70 Jurskis 2005.

71 Turner *et al.* 2008.

72 Stone *et al.* 2008.

73 University of Western Sydney 2013.

74 Bobbink *et al.* 2010.

75 Ostry *et al.* 2010.

76 White 1993.

77 White 1986.

78 Jurskis 2008.

79 White 1986.

80 Palzer 1981, Jurskis 2005.

81 Podger 1981, Manion 1991, Jurskis 2005.

82 Wardlaw 1989, Jurskis 2005.

83 Jurskis 2005, Close *et al.* 2009.

84 Manion 1991.

85 Auclair *et al.* 1992.

86 Auclair 2005, Fig. 4.

87 Manion 2003.

88 Jurskis 2009, 2011a.

89 Jurskis 2008.

90 Ellis 1964, Ellis *et al.* 1980, Ellis and Pennington 1992.

91 Close *et al.* 2011a.

92 Close *et al.* 2009.

93 op. cit.

94 Close *et al.* 2011b.

95 Hesterberg and Jurgensen 1972.

96 op. cit.

97 Horton *et al.* 2013.

98 Horton 2012.

99 Steinbauer *et al.* 2015.

100 Starck 2014.

101 University of Western Sydney nd.

102 Hawkesbury Institute for the Environment nd, University of Western Sydney nd.

103 Tozer and Keith 2012.

104 Pringle *et al.* 2009.

105 Ford *et al.* 2009.

106 Tasker and Dickman 2004.

107 NSWEPA nd.

108 McCallum 2008.

109 Jurskis 2004b.

110 Parks & Wildlife Service Tasmania 2014.

111 Pavlic *et al.* 2008, Scott *et al.* 2009, Taylor *et al.* 2009, Barber *et al.* 2013.

112 University of Western Sydney 2013.

113 McFadden 2013.

114 Hunter *et al.* 2009.

115 Merritt 2007.

116 Jurskis *et al.* 2003.

117 Worboys 2003.

118 McCallum 2008.

119 Australian Government Department of the Environment nd a.

120 Jones and Barmuta 2000.

121 Ibid.

122 Ibid.

123 Jurskis 2009.

124 Fisher 1983, Reid and Yan 2000, Miller *et al.* 2003.

125 Watson 2002.

126 Tench 1793.

127 Darwin 1845.

128 Watson 2009.

129 Watson 2002.

7

THE WILDERNESS MYTH

The only true wilderness is between a greenie's ears
New millenium bumper sticker

Wilderness is an elitist, imperialist, antisocial concept rooted in the arrogance of Terra Nullius – the idea of land belonging to no-one – which helped the Roman Empire to flourish and also allowed European aggressors to dispossess Aborigines around the world. Australian explorers were well aware that Terra Nullius had no application as Aborigines either tried to repel them, welcomed them to share resources or fled to avoid them. The High Court of Australia finally overturned this concept in the Mabo decision of 1992, but ecologists and environmental bureaucrats have mostly not caught on to the obvious implications for land management.

National Park was established south of Sydney in 1879 as a fairdinkum park – a place of public pleasure. It had clearings and gardens, infrastructure and facilities. There was a sawmill as well as gravel pits and clay pits to provide materials and revenue for construction and maintenance. After Her Majesty Queen Elizabeth II's visit in 1954 the park was renamed Royal National Park. It was only the second national park in the world after Yellowstone in the USA was established seven years earlier. Although Yellowstone was never officially declared a wilderness, its turbulent history of active mismanagement alternating with neglect under a wilderness philosophy makes interesting reading of Alston Chase's *Playing God in Yellowstone*.

Forester Aldo Leopold is generally credited with unleashing the concept of wilderness on modern society by persuading the United States Forest Service to designate New Mexico's Gila Primitive Area in 1924. However the Colong Foundation claims that Sydney based teacher of architecture, Myles Dunphy was the original guru of wilderness. Dunphy assembled an elite group of city-dwelling bushwalkers – the Mountain Trails Club – in 1914. In 1932 they formed the National Parks and Primitive Areas Council to lobby the NSW government to establish national parks as exclusive playgrounds for the privileged few with the health, wealth and spare time to take solace and refuge from society in uncivilised bush.

The Council adapted the philosophical ramblings of naturists such as Thoreau and Leopold to plant the seeds of forbidden fruit that wilderness represents to your average Adam and Eve. Thoreau famously declared "In Wildness is the preservation of the World". Dunphy and friends pushed the corollary – in human economy is the destruction of the world. By 1967, when the National Parks and Wildlife Act was passed in NSW, these seeds were bearing fruit, and the new parks authority was born with a mission to atone for the Original Sin of development.

As standards of living improved after World War II, wages increased and technology was substituted for rural labour. Population centralised in cities and the voting power of municipalities equalled that of rural regions. Governments focused on the wants of city-dwellers who had little economic or cultural stake in the bush. They were mostly concerned with recreation and 'preserving nature'. Blue Mountains National Park was secured as Dunphy's playground in 1959 and his Council was able to rest on its laurels after the National Parks and Wildlife Service was established. Then the Colong Foundation took on the mission for wilderness.

This Foundation denies that wilderness is based on Terra Nullius in the following terms:

> The wilderness movement knows that it has several things in common with the traditional Aboriginal relationship with the land including: … love of quiet contemplation of one's surroundings and an awareness of the spiritual quality of place. The conflict arises where Aborigines want to … manage the land in a way which interferes with natural processes.[1]

Denial of Aboriginal economy is, of course, central to Terra Nullius. But wilderness guru Aldo Leopold went even further and denied Aboriginal culture:

> Wilderness is the raw material out of which man has hammered the artefact called civilisation. Wilderness was never a homogenous raw material. It was very diverse, and the resulting artifacts are very diverse. These differences in the end-product are known as cultures. The rich diversity of the world's cultures reflects a corresponding diversity in the wilds that gave them birth.
>
> For the first time in the history of the human species, two changes are now impending. One is the exhaustion of wilderness in the more habitable portions of the globe. The other is the world-wide hybridisation of cultures through modern transport and industrialisation.[2]

In the late 1960s and 1970s, Australian and North American anthropologists rediscovered historical ecology and realised, as the explorers had, that wilderness (excepting Antarctica) and Terra Nullius were nonsense. They recognised Aboriginal cultures and economies. Henry T Lewis in *Before the Wilderness* noted that his was a second attempt to generate anthropological interest in Aboriginal burning and shared credit for the attempt with Australians Rhys Jones and Sylvia Hallam.[3] I'm indebted to my Washoe 'Indian' friend Darrel Cruz for pointing me to this book which provides an overview of recent anthropological awakening as well as the original dispossession of Native Americans and consequent environmental degradation in western North America.

I met Darrel in 2011 after he gave a fascinating presentation on the use of modern heavy machinery and engineering as well as traditional burning, hunting and gathering to restore natural species, landforms, hydrology and ecosystem function within his tribal lands around Lake Tahoe. This work provides an interesting contrast to the sorry history of Yellowstone where reintroductions and restorations have consistently failed.[4] The Washoe successfully reintroduced native fish that are important traditional foods. Darrel explained to me that the land and people are tied together in one ecosystem: "we are dependent on the land and the land is dependent on us to care for it".

Australian Aborigines have been equally eloquent in expressing this fact of life. Gagudju Man, Big Bill Neidjie put it poetically: "I telling you this because the land for us, never change round. Places for us, earth for us, star, moon, tree, animal. No-matter what sort of animal, Bird snake … all that animal like us. Our friend that".[5] Australian Living National Treasure Galarrwuy Yunupingu reportedly said that "an Aboriginal deprived of his tribal land is like a leaf torn from its tree".[6] These sentiments are entirely lost upon wilderness enthusiasts who have firstly used Aborigines to change land tenure and then tried to prevent them from using their regained lands.

For example, Tim Flannery lamented one example where Aboriginal ownership was restored:

> Even under Labor governments with a strong green bent, national parks are not always safe. In 2010 the Queensland Bligh government began the process of de-gazetting a large part of Mungkan Kaanju National Park on Cape York Peninsula with a view to giving the land back to its traditional Aboriginal owners.[7]

The incoming Liberal National government decided to repeal Wild Rivers legislation to allow traditional Aboriginal owners to manage this land according to their own wishes. Cape York Aboriginal leader Noel Pearson had said "I want to see whitefellas, blackfellas, greenfellas all working for a balanced future. … At the moment what we've had under the Wilderness Society and the Labor government is the greenfellas putting their foot on our throats".[8]

Flannery apparently considered that blackfellas couldn't be trusted to look after their country:

> In the same vein the Newman government looks set to repeal the Wild Rivers legislation, replacing it with a far north Queensland land use strategy whose details are yet to be determined. It's too soon to know the impact of such changes on species protection. But a softer line … on river health could have massively damaging effects.[9]

When the Whitlam Government abolished fees in 1974, universities increasingly attracted student activists looking for a cause rather than a career. The wilderness power base shifted from upper middle class bushwalkers to 'socialist' greens, and the Wilderness Society established in 1976. Three decades later it had about 2,500 members and an annual income of 13 million dollars. Thus from small, elite groups of bushwalkers and naturists has grown an army of feelgooders with hugely disproportionate political power compared to stakeholdings, and powerful representation within the bureaucracy and academia. Governments now spend large amounts compensating socioeconomic destruction and supporting inefficient mismanagement of reserves (high cost : benefit) for political advantage.

Wilderness hysteria erupted in Australia with the Franklin Dam debacle of 1983 in Tasmania. This created a tsunami of political and socioeconomic as well as ecological consequences. The political shenanigans included the 1983 World Heritage (Western Tasmania Wilderness) regulations and culminated in the High Court Case of Commonwealth versus Tasmania that affirmed the constitutional right of the new Hawke Labor government to meddle in States' affairs. Hawke promised to stop the dam, and gained power via mainland city electorates even though Tasmanian seats were all retained by the previous government now in opposition. The High Court decision was a four to three majority, with the Chief Justice on the minority, and former ALP parliamentary member Justice Lionel Murphy on the majority. Murphy had been controversially appointed to the High Court by Prime Minister Whitlam and controversially replaced in the senate by a non-affiliated appointee of NSW Liberal Premier Tom Lewis, allowing a supply bill to be blocked and paving the way for a constitutional crisis and Whitlam's dismissal.

In 1985 Murphy was convicted of attempting to pervert the course of justice by allegedly seeking to influence the Chief Magistrate of NSW, Clarrie Briese, in a matter connected with Murphy's "little mate" Morgan Ryan. After the conviction was overturned on appeal, he was retried and acquitted. In May 1986, Attorney General Lionel Bowen established legislation for a Parliamentary Committee of Inquiry into Murphy's fitness to remain on the High Court. In July, Murphy announced that he had terminal cancer, and he died in October.[10]

Murphy's attitude to use of natural resources and the rights of regional stakeholders was evident in his deliberations on Commonwealth versus Tasmania:

> The world's cultural and natural heritage is, of its own nature, part of Australia's external affairs. It is the heritage of Australians, as part of humanity, as well as the heritage of those where the various items happen to be. As soon as it is accepted that the Tasmanian wilderness area is part of world heritage, it follows that its preservation as well as being an internal affair, is part of Australia's external affairs.[11]

As a consequence of this philosophy, Commonwealth environmental legislation can now be used as a sledgehammer to crack nuts. For example, minor trials of traditional alpine grazing to reduce fire hazards in Victoria, and continuation of some small scale selective logging operations in south-eastern NSW were blocked by Federal Environment Minister Tony Burke. Also, proposed micro-scale thinning trials in forests that didn't exist before European settlement, require assessments and approvals under Commonwealth environmental legislation (see Chapter 11).

The first specific wilderness legislation in Australia was the NSW Wilderness Act of 1987. There is still no specific Commonwealth legislation, but wilderness enthusiasts are empowered by the 1992 National Forest Policy Statement. Any fair reading of the definitions in these documents will demonstrate that wilderness of any substantial area, and the area must be substantial to satisfy the definitions, cannot exist anywhere other than the unpeopled continent of Antarctica.

The NFPS definition of wilderness – "not substantially modified by and remote from the influences of European settlement"[12] – excludes virtually the whole of Australia as a result of post-European impacts on fire regimes and consequent changes in vegetation.[13] In NSW, the

Wilderness Act 1987 provides that areas may not be identified as wilderness unless they have not been substantially modified by humans or can be returned to such a state. Unfortunately the test of this provision is the opinion of the Director-General, so it's a bit like setting the fox to mind the chooks. But it's quite obvious that we can't resurrect the megafauna and the environment which was irrevocably altered by the arrival of Aborigines despite Flannery's curious suggestion to introduce the Komodo dragon as a replacement megacarnivore.[14]

The rarity of areas meeting the criteria for wilderness is evident in the case of the Wollemi pine. This living fossil survived human arrival and European settlement because it grows on narrow shelves in deep, narrow, sheltered gorges that were totally inaccessible to man and his firestick[15] until a wilderness enthusiast with plenty of spare time and presumably the best modern abseiling equipment, chanced upon it in 1994. This true wilderness however, does not meet the size criterion of the Wilderness Act. There are about 100 mostly clonal trees in a few small stands.[16] It is ironic that naturists are responsible for the real threat that the species will become extinct like most other fossils, because they introduced phytophthora to the mini wilderness, causing disease in at least one of the stands by 2005.[17] In this case, protecting a real wilderness, however small, rather than creating a playground for the elite would have effected conservation of a truly endangered species that had endured through several glacial cycles. Unlike the megafauna, Wollemi pine survived human economy for forty or fifty thousand years because it occupied the ultimate refuge.

William Cronon wrote that:

> Wilderness hides its unnaturalness behind a mask that is all the more beguiling because it seems so natural. As we gaze into the mirror it holds up for us, we too easily imagine that what we behold is Nature when in fact we see the reflection of our own unexamined longings and desires. For this reason, we mistake ourselves when we suppose that wilderness can be the solution to our culture's problematic relationships with the nonhuman world, for wilderness is itself no small part of the problem.[18]

In an essay on *The paradoxes of wilderness fire* Stephen Pyne wrote "The premise behind wilderness is that it stands outside culture. It can't. Rather, there are compelling arguments for wilderness as a state of mind more than a state of nature and for the wilderness idea as an invention".[19] As we have seen, it was an invention whose time had truly come after Whitlam promoted a subculture of student activists looking for a cause which they found in the Franklin Dam proposal. Hawke seized the opportunity and empowered them, creating the tidal wave of the Greens' political power under Bob Brown, which culminated after three decades in the disastrous minority Gillard/Brown government. Brown ended his political career as the proverbial rat jumping ship before it sank in 2013.

Five years earlier, I had the great pleasure of attending a public meeting at Bermagui where Brown had the star billing. He had been invited by local anti-logging activists to look at 'forest destruction'. Brown was obviously underwhelmed by the outcomes of the very selective logging[XI] that he saw, and he focused instead on the 25th anniversary of the saving of the Franklin. This was at a time when the Greens were demanding a carbon tax to limit consumption of fossil fuels and promote renewable energy. The zeal of those that would nobble Australia's only proven, viable and sustainable renewable energy industry for the sake of a self-contradictory mythology was truly wondrous to behold. On the 30th Anniversary of the 'saving', ABC radio ran a story promoting renewable energy followed, ironically, by a story about a celebratory dinner where Hawke praised Brown but cautioned that Greens don't pay enough attention to economics.

When Brown formed a minority government with the ALP and some egotistical independents, voters saw Prime Minister elect – Julia Gillard's – lust for status overcome any consideration of Australia's socioeconomic wellbeing, and they saw the destructive potential of wilderness enthusiasts with power. The receding wave of the wilderness tsunami will hopefully take much debris, such as Green political influence, back to the depths

---

XI Trees are selected for removal or retention based on their value for timber and their environmental values.

where it belongs. Unfortunately, Tasmania's socioeconomic fabric is amongst the detritus being sucked away. The final dismantling of its formerly thriving native forestry industry in 2012 left the State destitute and, given its spectacular landscapes and rich cultural heritage, disproved beyond doubt the green myth that tourism can sustain an economy without substantial primary and/or secondary industry. Additionally, the 2013 east coast megafires and the looming extinction of the Tasmanian devil show that the consequences of embracing wilderness mythology were environmentally as much as socioeconomically disastrous.

Endnotes

1 The Colong Foundation for Wilderness Ltd. 2012.

2 Leopold 1948, p. 188.

3 Lewis 1993.

4 Chase 1987.

5 Neidjie 1989.

6 Flood 2006.

7 Flannery 2012.

8 Ryan 2011.

9 op. cit.

10 National Archives of Australia 2015.

11 Murphy 1983.

12 Commonwealth of Australia 1995.

13 Mitchell 1848; Howitt 1891; Jurskis 2009, 2011; Gammage 2011.

14 op. cit.

15 Australian Government Department of the Environment nd b.

16 Ibid.

17 Ibid.

18 Cronon 1996.

19 Stephen Pyne pers. comm. 2015.

8

HUMAN FIRES, MEGAFIRES, PARAMILITARIES AND GLOBAL WARMING ENTHUSIASTS

They ran their heads very hard against wrong ideas, and persisted in trying to fit the circumstances to the ideas, instead of trying to extract ideas from the circumstances

Pip in Charles Dickens' *Great Expectations*

First-fleeter, Watkin Tench left no room for doubt about the role of human fire in Australia's natural environment:[1]

> The country, I am of opinion, would abound with birds, did not the natives, by perpetually setting fire to the grass and bushes, destroy the greater part of the nests; a cause which also contributes to render small quadrupeds scarce.

His account of an early exploration near Port Jackson in April 1791 gives an insight into the nature of the landscape under Aboriginal fire management and how it changed as a consequence of European settlement.

> Every man (the governor excepted) carried his own knapsack, which contained provisions for ten days. If to this be added a gun, a blanket, and a canteen, the weight will fall nothing short of forty pounds. Slung to the knapsack are the cooking kettle and the hatchet …
>
> … The country for the first two miles, while we walked to the northeast, was good, full of grass and without rock or underwood. Afterwards it grew very bad, being full of steep barren rocks, over which we were compelled to clamber for seven miles.

The place where the country "grew very bad" was subsequently called North Rocks. Now, after two centuries of post-European fire suppression, the steep, formerly barren rocks are barely visible for dense, sometimes impenetrable scrub. But in 1791, during his six-day excursion, Tench made only a single reference to difficulties occasioned by vegetation:

> Our natives continued to hold out stoutly. The hindrances to walking by the river side which plagued and entangled us so much, seemed not to be heeded by them, and they wound through them with ease; but to us they were intolerably tiresome. Our perplexities afforded them an inexhaustible fund of merriment and derision: did the sufferer, stung at once with nettles and ridicule, and shaken nigh to death by his fall, use any angry expression to them, they retorted in a moment.

After leaving the river and returning to rocky sandstone country with "nothing but trees growing on precipices" Tench "saw a tree on fire here and several other vestiges of the natives". Scrambling to the top of a sandstone mountain "with infinite toil and difficulty", Tench had an extensive view "in almost every direction, for many miles". Governor Phillip named this "pile of desolation" Tench's Prospect Mount. Tench's "infinite toil and difficulty" would pale into insignificance against the task that would confront him on this scrubby hill today, and the toil would be in vain since his distant objective, Richmond Hill, is no longer visible because it is obscured by thick scrub.

This event featured in Eleanor Dark's 1941 classic historical novel *The Timeless Land* which was, for a time, prescribed reading for secondary school students. It was also included in Tim Flannery's 1998 volume *The Explorers*. When I was researching changes in fire regimes and vegetation on the sandstone country around Sydney,[2] I was naturally curious to visit the Mount and see how it had changed. I was astonished to find that the location of such an historic site was unknown. It wasn't mapped, or recognised by the Geographical Names Board (GNB). I compared Tench's journal against a modern topographic map and worked out the location. Tench's journal contained a map that was not consistent with the text nor with an independent account of the journey published by

Captain John Hunter. Reliance on the erroneous map had caused much confusion and debate within the Royal Australian Historical Society (RAHS) in the 1920s and 1930s. The correct location has now been published in the Journal of RAHS[3] and the GNB is still deliberating.

Mitchell's observations of Aboriginal fire management were paraphrased in Chapter 3. He observed that "no vegetable soil is formed" on sandstone country because "conflagrations take place so frequently and extensively ... as to leave very little vegetable matter to return to earth".[4] (A modern ecologist who reviewed a paper that I developed with Roger Underwood,[5] disparaged these observations as "Mitchell's erroneous inferences".)

Travelling the Great Northern Road, Mitchell described the country between Parramatta and the Hawkesbury River thus:

> no objects met the eye except barren sandstone rocks, and stunted trees. With the *banksia* and *xanthorhaea* always in sight ... The horizon is flat ... the whole face of the country is composed of sandstone rock, and but partially covered with vegetation.[6]

He surveyed the continuation of the road north of the river, and recounted that:

> on many a dark night ... I have proceeded on horseback amongst these steep and rocky ranges, my path being guided by two young boys belonging to the tribe who ran cheerfully before my horse, alternately tearing off the stringy bark which served for torches, and setting fire to the grass trees (*xanthorhaea*) to light my way.[7]

This sandstone country is now so densely shrubbed that it is possible neither to run through it nor to ride a horse through it nor to view the horizon, even in broad daylight. Furthermore the old stringybarks and grasstrees have succumbed to megafires and disease. The immature stringybarks have tight bark that cannot be quickly stripped by hand for torches, and the few young grasstrees have virtually no trunk nor skirt of dry leaves that could be lit for illumination.

On 10th and 11th February 1791, Aboriginal fires in woodlands on sandstone country northwest of Rose Hill (Parramatta) were burning under extreme temperatures (> 43^0 C) and scorching northwesterly winds. In what is now Parramatta Park, parrots and flying foxes were dropping dead from the trees. But the European settlement was not affected by the fires.[8]

On another blow up day,[XII] 5th December 1792, a grass fire at Sydney burnt one house and several gardens and fences before being controlled. On the same day Collins wrote:

> At Parramatta and Toongabbe also the heat was extreme; the country there too was everywhere in flames. Mr. Arndell was a great sufferer by it. The fire had spread to his farm; but by the efforts of his own people and the neighbouring settlers it was got under, and its progress supposed to be effectually checked, when an unlucky spark from a tree, which had been on fire to the topmost branch, flying upon the thatch of the hut where his people lived, it blazed out; the hut with all the outbuildings, and thirty bushels of wheat, just got into a stack, were in a few minutes destroyed.[9]

Though burning under extreme conditions, these fires were obviously in light, discontinuous fuels, and did not develop into firestorms. But in January 1994, fires burning under similar weather conditions,[10] in dense, three dimensionally continuous fuels produced by post-European neglect, were mostly uncontrollable, burning more than thirty thousand hectares around Sydney, and claiming hundreds of houses and three lives despite the efforts of a well-equipped army of firefighters. The difference was due to the development of fire storms producing showers of embers and long distance spotting. For example, the run of fire that claimed human lives spotted 800 metres across a major watercourse.[11] However there were four localities where runs of fire during extreme weather were contained in areas that had prior hazard reduction burning.[12]

Matthew Flinders saw true wilderness, thick forests and the legacy of megafire on Kangaroo Island. When he approached the mainland after leaving the Island his journal records that:

XII A day when strong winds, low humidity and high temperatures can promote extreme fire behaviour.

> Fires were seen ahead; ... At daylight I recognised Mount Lofty
> ... The nearest part of the coast ... mostly low, and composed
> of sand and rock, with a few small trees scattered over it; but at
> a few miles inland, where the back mountains rise, the country
> was well clothed with forest timber [woodland], and had a fertile
> appearance. The fires bespoke this to be a part of the continent[11]
> [inhabited by man].

Edward Curr described Aboriginal management as fire-stick farming, and Mitchell eloquently portrayed the interdependence of fire, grass, grazing animals and man.[13] Aboriginal burning killed seedlings of trees and shrubs in all environments that were physically accessible to low and moderate-intensity fires. These fires maintained nutrient cycling, healthy ancient trees and ancient shrubs, herbs and grasses, bare patches, animals adapted to these habitats, and the regeneration potential of the dominant trees and shrubs, whether scattered individuals in woodlands or open stands in forests. The natural landscape had low levels of discontinuous fuels and no megafires. People travelled easily across the landscape and were able to see far ahead, except where physical refuges from fire allowed dense vegetation to develop.

Mitchell, Curr and Howitt[14] recognised that woody thickening, megafires and chronic eucalypt decline were consequences of surrendering human firepower. By 1851 fire suppression had created a landscape bomb of unnaturally dense and shrubby forests that exploded into fire on Black Thursday 6th February[15] and burnt five million hectares of Victoria. By ignoring the historical lessons, foresters trained in Europe doomed us to repeat the disastrous mistakes during the early-twentieth century.[16] After another generation of foresters learnt from past mistakes, firestick forestry was applied for only a couple of decades from the middle of the century,[17] until antisocial naturists with wilderness between the ears swept to political power on a destructive tsunami. Ongoing exclusion of fire across broad areas has recreated and maintained a landscape that explodes into megafires at an average interval of eleven years across much of southern Australia.[18]

In recent decades, Australia has wasted billions of dollars on research, emergency response capacity, and rebuilding and repairing after megafires consequent to a wilderness mentality and ignorance of firestick ecology. Absurdly, an advocate of fire suppression, Professor David Lindenmayer, claims that logging has caused this problem.[19] In fact, the impacts of logging on forest structure have been miniscule compared to the impacts of repeated megafires caused by exclusion and suppression of fire.[20] The prudent course of action would use logging to provide infrastructure and resources to support broadscale burning across the landscape.

In *The Still-Burning Bush*, fire historian Stephen Pyne poetically summarised forestry's place in Australia's ecological history:

> Australian forestry broke away from the global trend and, after chastising rural Australia for its promiscuous burning off, similarly adapted the firestick, but brought to its use the intellectual discipline of science and the administrative discipline of state-sponsored forestry. This was a singular achievement, unrivalled elsewhere in the developed world, and as those other industrial nations wrestled with the consequences of fire's abrupt removal, they saw in Australia an expression of what they had lost and what they might wish to reclaim. Call that achievement a firestick forestry.
>
> Over the past few decades, however, that synthesis has disintegrated. State-sponsored forestry has imploded, surrendering rural fire protection to an emergency-services bureaucracy and yielding public land management to nature reserves. This time – so far – no one has seized the firestick and redirected it to new purposes, what one might loosely categorise as firestick ecology. Instead the country seems locked into a polarised politics of identity for which conflagrations provide a dramatic backdrop.[21]

Open, productive and fire-safe stands of mallee, cypress and river red gum have been replaced in nature reserves by dense, even-aged, homogenous scrubs that are literally time bombs and virtually biological deserts. Many foresters in Victoria think that mountain ash and alpine ash are sensitive to fire because they now mostly grow in scrubby stands that are inaccessible to mild fires and therefore vulnerable to severe fires. Thus public land managers have locked themselves into a vicious cycle of hopeless attempts at fire suppression, and inevitable megafires.

The Black Friday fires started on 13th January 1939, incinerating two million hectares of Victoria, and taking 71 lives.[22] Stretton's subsequent Royal Commission reported that the main cause of catastrophe was the dense shrubby condition of the forests because burning by the Forests Commission was "ridiculously inadequate".[23] But Phil Cheney informs me that historical records show a large number of small fires in far East Gippsland where graziers had not been subdued by the Forestry Commission. This was recently verified independently by John Mulligan who experienced these fires as a boy:

> Even the 1939 fires did not take hold in the far east even though we had fires on that terrible day they did not have the available fuel to take hold. I was there, I remember that day. And this was because the "No Burn" policies had not yet been able to be policed in the far east.[24]

However, megafires devastated a large proportion of Hume (upper Murray) and Snowy River Catchments in New South Wales where there were proposals for a large scale river regulation and diversion scheme to provide hydroelectricity and irrigation water whilst controlling floods. The history of Hume-Snowy Bushfire Prevention Scheme, established in 1951 in response to lessons such as those of Black Friday, provides a cameo of firestick forestry's rise and fall in south-eastern Australia, and its eventual subjugation to paramilitary madness.

Around the late-nineteenth and early-twentieth centuries, foresters gazetted timber reserves, and silviculturally treated alpine ash stands. Concerned about damage to timber resources, they attempted to restrict burning by graziers. In 1925 the River Murray Commission also raised concerns about erosion attributed to grazing and deliberate burning. Extensive wildfires affected most of the Alpine Region in 1926 and 1939 destroying nearly all the Victorian alpine ash stands and causing severe erosion. However some stands that had been illegally burnt by graziers escaped serious damage.[25]

Forester Baldur Unwin Byles surveyed the Hume catchment in 1931/32 and raised concerns about vegetation changes and erosion caused by fires. Byles claimed that "99% of fires" were a result of stockmen burning rank grass.[26] This may well have been the case. If so, the ecological changes and severe erosion that he was concerned about were obviously caused by only 1% of fires – the wild ones. Byles initially recommended fire suppression together with increased grazing to prevent the accumulation of rank grass. After he was appointed as forestry representative on Kosciusko State Park Trust, other trustees apparently convinced him that grazing also caused serious erosion and therefore should also be eliminated.[27]

Kosciusko State Park in NSW was gazetted in 1944, and by the time the Snowy Mountains Scheme commenced in 1949, 12,000 hectares had been withdrawn from grazing.[28] At this time Alec Costin carried out investigations for the Soil Conservation Service of NSW (SCS) and claimed that burning by graziers had destroyed woodlands, triggered insect plagues that were destroying grasslands, promoted shrub invasion and caused massive erosion. His prejudice against prescribed burning was such that he suggested using persistent organochlorine insecticides to control the insects that had irrupted after burning was prohibited.[29]

Meanwhile foresters had, in the school of hard knocks, learned lessons that should have been obvious from the historical ecological narratives of naturalists such as Mitchell, Curr and Howitt as well from the history of wildfires after Aboriginal burning was disrupted. Megafires, severe erosion, irruptions of pests, eucalypt decline and scrub invasion had occurred in the Alpine Region, as in other regions, as a consequence of attempted fire suppression. Foresters aligned with graziers and bushfire brigades to set up the Hume-Snowy Bushfire Prevention Scheme which aimed to protect lives, infrastructure and natural resources across eight hundred thousand hectares of New South Wales by hazard reduction burning. Unfortunately, foresters under Byles' influence still thought that they could protect timber values by excluding fire from alpine ash stands.

Right from the start there was tension between proactive fire managers and conservation authorities who were generally opposed to grazing or burning. Some of this tension may have had an element of 'sour grapes'. In 2006 Australia's Academy of Science published a lengthy interview with Costin[30] wherein he stated his reasons for starting as a cadet with the SCS:

> My main reason was that if I wanted any tertiary education I would have to get some support, and the most attractive support took the form of the very few Public Service cadetships sponsored by some of the main government departments in Sydney – Agriculture, Forestry, Water, Public Works et cetera. I was advised to have a go at one of these cadetships, and I applied for one in forestry. In a subsequent interview as one of the people that were short listed, I was told that I'd missed out on the forestry cadetship (subsequently I've been very happy about that). ... then Sam Clayton, the Soil Conservation Commissioner, told me there was a cadetship available in agriculture but from the Soil Conservation Service of New South Wales. That's how I started my university course in agricultural science.

Severe fire seasons occurred in Hume-Snowy during 1951/52 when sixty fires burnt about fifty thousand hectares and 1965 when sixty six thousand hectares were burnt.[31] After a fire burnt from the Tumut River through the north of the Park in 1965, graziers criticised the Hume-Snowy bushfire authorities for a tardy response and suggested that the fire was not controlled until it ran into country that was still grazed under lease, where lessees stopped it in light fuels.[32] But Phil Cheney says that the front died out in the morning dew in grazed native pastures, whilst most of the perimeter in the forest country was controlled by the Forestry Commission under Bob Richmond before rain finally finished it off.[33]

Another severe fire season in 1968 prompted a reassessment of the fire management strategy by eminent fire scientist Alan McArthur and then chief of the Hume-Snowy Bushfire Scheme, Roy Free. More than 10% of the Hume-Snowy Fire District had been burnt by wildfires between 1965 and 1968. McArthur and Free considered that the existing fire protection arrangements were not capable of keeping damage to regional values at an acceptably low level because fuel reduction had only been carried out along roads and fire trails.[34]

By this time, a new technology of aerial burning had been developed for jarrah forests in Western Australia. McArthur and Free noted that large areas could be treated, under very precise prescriptions, at low cost, taking advantage of very small windows of opportunity in terms of seasonal conditions. They recognised that exclusion of fires from sensitive areas such as subalpine bogs could only be practically achieved by burning in the surrounding landscape at times when these areas were not flammable. A trial burn of 5,000 ha in the rugged Flea Creek catchment west of the Australian Capital Territory in April 1967 had caused minimal damage and less than 5% crown scorch. McArthur and Free initiated a new plan for fire protection in Hume-Snowy including strategic burning in the south and southwest, fuel reduction right through the forests, and protection of small areas of special value using burning by ground crews. They proposed aerial burning in autumn at about five year intervals and ground burning of special value areas at about three year intervals. They considered that about 60,000 hectares should be burnt each year, comprising about 25,000 ha strategic and 35,000 ha general fuel reduction burning.[35]

At the same time, Dr Graham Edgar was appointed by Premier Tom Lewis to conduct an independent inquiry into grazing in the Park. Edgar's report in 1969[36] simply restated Costin's claims that grazing and burning caused erosion.[37] He recommended to the Minister for Lands that grazing be prohibited, that freehold properties be acquired and that the Government take responsibility for controlling rabbits and blackberries. Edgar noted that hot fires within the preceding decade caused much destruction of snow gum stands in the park,[38] but he didn't distinguish between the impacts of prescribed burns and wildfires. The Snow Lessees and Occupiers Association pointed to many shortcomings in his report and in the approach of Government Agencies to land management in the region:

> There was no consideration of data showing better water yields from grazed catchments. The severe impacts of wildfires compared

to grazing were not considered. Shrub invasion, wrongly attributed to grazing, was promoted by fire exclusion, and [it] increased fire hazards in 'protected' catchments. Costs of pest and weed control,[XIII] formerly borne by graziers were being transferred to taxpayers.[39]

Another extreme fire season occurred in 1972/73 after a dry winter. There were 25 days of very high and extreme fire danger including three days of severe lightning storms that started many fires. Forty nine wildfires burnt 68,000 ha including 48,000 ha in Kosciusko National Park. No substantial damage occurred to assets or infrastructure, and there was no loss of life or even serious injury. The cost of firefighting operations in Hume-Snowy was relatively low compared to other fire districts that faced similar circumstances. But some catchment areas were damaged by high-intensity fires where scheduled prescribed burns had been cancelled in response to concerns by Commonwealth Scientific and Industrial Research Organisation (CSIRO), the Soil Conservation Service and the National Parks and Wildlife Service (NPWS).[40]

Three large fires burnt in the park. The Welumba Creek fire of 4000 ha was controlled by limited manpower aided by the fuel reduction benefits of the strategic burning program. The Jacobs River fire of 30,000 ha escaped control in an area that had been scheduled for aerial burning in the previous year but was withdrawn from the program following objections from CSIRO, SCS and NPWS. The Byadbo section of the Park had been burnt the previous year and presented no problems. Control of the Grey Mare fire of 13,000 ha was hampered by a lack of fuel reduction burning in alpine ash stands. As a result it was suggested that prescribed burning should be routinely carried out in these stands using more sophisticated prescriptions, closer monitoring of fuels and weather, and more precise ignition (from helicopters) than in the general burning program.[41]

XIII Rabbits and blackberries were still running rampant in 2013, particularly around former freehold properties that are now used as depots by the parks service. Brumbies were also running rampant, but unlike rabbits and blackberries, they appeared to be keeping some areas in a more natural, open condition despite the lack of prescribed burning.

In fact, prescribed burning was reduced after 1974 when a new National Parks and Wildlife Act was passed. Ecologists from CSIRO, SCS and NPWS increasingly influenced land management, and the Scheme was finally disbanded in 1986.[42] Severe wildfires in 1978, 1983 and 1988 burnt similar areas as previous fires, but at much higher intensities, causing substantial soil erosion, destroying high quality timber resources, and causing large economic losses to neighbouring graziers.[43]

In 2001 the NSW Government determined to review the plan of management for Kosciuszko National Park, and appointed an Independent Scientific Committee (ISC) to assess the values of the park. The ISC had a bizarre view of its fire management history:

> Fire management in the park has developed … from a very simplistic approach … with an annual program of fuel reduction to an approach based on ecological principles that provide for sound nature conservation, catchment stability and the maintenance of an acceptable level of risk from wildfire impacts on infrastructure, neighbours and park users. … The Aboriginal tribes … did light fires … but … Only small areas would have been burnt … the [Hume-Snowy] fuel reduction program never reduced the fire hazard if one ever existed.[44]

When Deputy Surveyor-General Townsend explored the Alps in 1846, he noted that Aborigines had recently been gathering bogong moths and consequently "the country throughout the whole survey was burnt, leaving my bullocks destitute of food". Townsend referred to alpine ash in these terms: "There are some small portions of forest in which grows the large Eucalyptus tree known as 'Black',".[45] In the natural system, mild annual fires burnt the grass, blackened the fibrous bark on the trunks of alpine ash and maintained a suitable environment in which seedlings could establish when an old ash tree burnt down or was blown over. Dense even-aged stands and severe landslips are a consequence of post-European megafires. Fig. 6 (p. 172) shows the bleached skeleton of an old 'black gum' that perished in the 2003 holocaust.

The Independent Scientific Committee attributed to prescribed burning, the erosion, woody thickening and shrub development caused

Fig. 6 Mature alpine ash killed by high-intensity fire in unnaturally heavy fuel, Kosciuszko 2003. (Photo by author)

by wildfires. They reported that "the number of trunks per tree has increased in regularly burnt areas, from one to three trunks per tree up to 40, with the number of trees per hectare having increased from under 50 to 6000-7000 per hectare".[46] In effect, they claimed that graziers and foresters had been using high-intensity crown fires. Also they claimed that dangerous fuel loads of up to 25 tonnes per hectare of "litter etc." were required to protect slopes from erosion. Furthermore they stated that trees such as snow gum and alpine ash needed to be protected from burning more often than once a century, and prescribed burning should only be carried out in "approximately 7%" of the park.

After the 2003 alpine megafires, a supplement to the ISC report explained that multiple lightning ignitions, which eventually culminated in Canberra's fire disaster, were unexpected. In fact multiple lightning ignitions during severe fire seasons have occurred repeatedly over millennia, but have never before had such severe consequences as in

2003. Multiple lightning strikes in State forests and private lands to the west of the park at the same time caused little damage, and were all controlled within three days.[47]

The ISC report and supplement illustrate the danger inherent in following the advice of scientists such as Costin who have no relevant expertise or experience in fire management. Costin said that:

> hazard reduction burning ignores one of the most fundamental resource equations that exist with respect to our natural and near-natural lands: fuel = catchment protection = habitat. ... a great deal of hazard reduction burning ... is absolutely inappropriate in most of our back country. Firstly, it never achieves what it is supposed to achieve, the reduced occurrence of devastating wildfires such as we've had. They are absolutely under the control of prevailing meteorological conditions combined with a preceding year or more of incredibly dry weather. If any fire starts under those conditions, let alone fires that are started by 20 or 30 or 50 almost simultaneous lightning strikes, there is no chance of doing anything with it until either it burns itself out – in other words, there is no more to burn – or there is a change in weather conditions. This has been the history of wildfires in Australia and it will continue. So some of these areas are going to get burnt anyway, but at nowhere near the frequency with which they are being burnt at the present time.[48]

The 2003 fires caused loss of human life, unprecedented erosion and siltation of water supply catchments, killed many rare and endangered plants and animals, destroyed hundreds of houses, thousands of stock, thousands of kilometres of fencing and tens of millions of dollars in public infrastructure. The vast majority of snow gum woodlands were reset to age zero. Taken together, the wildfires in 1978, 1983, 1988 and 2003 damaged or destroyed most of the habitat of the critically endangered corroboree frog. There is no doubt that they were the primary cause of the severe decline that threatens to extinguish this frog in the wild.[49] There is also no doubt that the megafires were a consequence of human fire management, not climate change as asserted by green academics and bureaucrats, because surrounding well-managed lands under the same climate did not suffer extreme fires despite multiple ignitions at the same time.[50]

The 2003 firefighting operations in NSW involved 1600 people, 36 dozers and 40 aircraft.[51] It is clear that general exclusion of fire with limited burning around assets cannot produce acceptable fire management outcomes even with huge financial and technological investments in fire suppression. A House of Representatives Select Committee inquiry into the 2003 megafires found that:

> The devastating loss of life, stock and property, the heart-breaking loss of bushland and wildlife, together with the tragic loss of confidence suffered by those directly affected by the bushfires, left a nation charred to its physical and spiritual core.[52]

But in 2006, Costin still stood by his advice that had helped to create this deplorable situation.

> A lot of the ecological communities that we are seeing now are communities that need 100 or 150 years' protection from fire. If you look at the demography of communities in the forests, say in snowgum woodland, you will find that possibly less than 5 percent of the entire community is old growth forest. The rest of it is succession forest in different stages – some burnt a few years ago, some from 1926 fires, '39 fires and so on. But the occurrence of old growth stands is diminishing. Simply, that is not because we're not having big wildfires [sic]. We are having a few of them, but if the area is big enough then the statistics are on your side. If you put fires through again and again and again, the statistics are *not* on your side. So we are losing the very naturalness which is one of the primary objectives in all our national park legislation.[53]

As well as the alpine fires across three million hectares, the Parliamentary Select Committee also inquired into fires covering another two million hectares in other regions.[54] It sat in many regional centers so that rural communities were well represented in evidence and in the report's recommendations. Chairman Gary Nairn reported that "The committee heard a consistent message right around Australia:- there has been grossly inadequate hazard reduction burning on public lands for far too long; local knowledge and experience is being ignored by an increasingly top heavy bureaucracy".[55]

The committee's recommendations addressed "Land management factors contributing to the severity of recent bushfire damage". Its first recommendation was to establish a national database of fuel loads and rates of accumulation according to vegetation type, climate and tenure. Its second was that "the Council of Australian Governments ensures that states and territories have adequate controls to ensure that local governments implement required fuel management standards on private property and land under their control".[56]

A dissenting report by a city-based Green member relied heavily on a submission from one of his constituents, Professor Robert Whelan, who had written a book on fire ecology despite lacking experience in management of land and fire. Whelan asserted that "broad scale hazard reduction is threatening biodiversity conservation and must therefore be avoided by land managers and resisted at a political level". The Council of Australian Governments (COAG) receives advice directly from the top-heavy bureaucracy criticised by Nairn and was therefore disinclined to adopt his recommendations. Instead they employed Whelan together with Professor of Forestry Peter Kanowski and former Special Forces Officer Stuart Ellis, none of whom have experience in forest management and prescribed burning, to conduct another inquiry. These men decided that education – "learning to live with bushfire" – was more important than reducing hazards; they claimed that we don't really know how prescribed burning affects biodiversity across the landscape; and they added insult to injury by questioning the effectiveness of hazard reduction burning.[57]

In the decade after the COAG Report of 2004 a further 186 human lives were lost, 173 of these in Victoria's Black Saturday fires on 7th February 2009.[58] Untold environmental damage has been caused by megafires that were a consequence of unscientific and antisocial green theories and philosophies. Despite overwhelming evidence that prescribed burning maintains ecosystem health, resilience and biodiversity as well as fire safety,[59] green academics and bureaucrats still don't 'get it'. Fire ecologists such as Professor Ross Bradstock argue that megafires are not unnatural, and are not reducing biodiversity. Indeed, they claim that high-intensity wildfires can increase biodiversity.[60]

So there is a dichotomy in Australia between a pragmatic, scientific and socially responsible approach to fire management based on history, experience and research of soundly based hypotheses on the one hand, and a theoretical, green, academic approach on the other. The scene was set in 1976 when the international Scientific Committee on Problems of the Environment – a trendsetter for contrived acronyms – initiated a project on *The Effects of Fire on Ecological Systems*. Under this project CSIRO scientist Dr. A. Malcolm Gill popularised the notion that the response of non-resprouting plant species to fire was either reproduction from seed or local extinction.[61]

We have since seen three decades of wasteful research around functional traits of plants and life cycles of obligate seeders[XIV] in an effort to specify minimum intervals between burning that will supposedly maintain biodiversity.[62] This is all based on false assumptions that prescribed fires reset vegetation to ground level or age zero, that ecosystems relatively unaffected by post-European economy are close to natural, and that maintaining these supposedly natural ecosystems maximises biodiversity. These false assumptions stem from ignorance of our ecological history and have been contradicted by many studies of prescribed burning.[63]

Some academics now realise they've reached a dead end with this theoretical approach. But they still want to do more research rather than letting land managers do the burning that urgently needs to be done. Gill and Stephens suggested that:[64]

> The development of trait-based approaches to fire management has possibly reached an impasse that cannot be breached without acquiring a much larger body of empirical data or a change in

---

XIV Plants that cannot resprout after they are scorched by fire, and therefore rely on seed to regenerate.

approach, such as classifying species' characteristics that determine responses to all components of the fire regimes, not just interval.

Governments will continue to support such inconsequential research because they like to be seen to be responding to disasters, and the disasters will continue while ever we're doing research instead of burning. But their biggest response has been the growth of massive paramilitary firefighting capacities including aircraft, trucks, command centres and sophisticated telecommunication facilities to ensure that their responses are widely seen. This has been at the expense of sensible land management. Local knowledge and experience usually doesn't receive due recognition in the command and control structure during firefighting operations, and the problems have been exacerbated by governments' need for Green support to retain power. For example, I will outline developments after lives were lost, hundreds of houses were destroyed and 24,000 people were evacuated during the 1994 fire disasters around Sydney.

Under New South Wales' *Bushfires Act* of 1949, as amended after each major fire event up to and including 1994, fire management on private, leasehold and vacant crown lands was the responsibility of volunteers under a very lean bureaucracy supported by local government. It was coordinated across the State by the Bush Fire Council in Sydney including ex-officio members from public land management and other agencies. Naturally, equipment, training, performance and coordination between agencies varied widely across the State. Assistant Forestry Commissioner Roy Free and Fire Management Officer Bob 'Tiger' Richmond were ex-officio Councilors and had been working on thorough reform of the legislation to improve overall standards of fire management and hazard reduction. The Coalition Government was receptive and the 1994 disaster gave impetus to the process.

Fahey's Coalition Government was defeated by Carr in 1995 and the proposed reforms were corrupted by green influence. The Rural Fires Act of 1997 came with a proviso that the objects relating to fire management were to be pursued "having regard to the principles of ecologically sustainable development". A peak green organisation, the Nature Conservation Council (NCC), was given a privileged position on the

Bush Fire Coordinating Committee, ensuring that these principles would be misconstrued, and would take precedence over fire management. In announcing the reforms, Rural Fire Service (RFS) Commissioner Philip Koperberg, a strong advocate of hazard reduction burning, pledged that they would make it easier to carry out burns.

In fact, NCC, NPWS and presumably RFS bureaucrats, worked to ensure that it would be more difficult by insinuating the untenable life cycles theory into the planning and approvals process. Now "acceptable" intervals between prescribed burns are those that will ensure environmental degradation and dangerous fuel loads.[65] This suits cost-squeezed public land managers, because everyday costs of management are reduced and firefighting is externally funded as disaster response. As Phil Cheney put it to me:

> The irony is that in the process of cost saving on government land they have shifted the costs to the suppression industry and created a monster that I fear, like in the USA, is unstoppable. Unless government land managers are made responsible for all suppression costs on their land then there is absolutely no incentive to undertake management that will reduce the hazards and make suppression cheaper and easier. The volunteers are caught up in this mess and are quite cynically used to avoid criticism of government actions. As Phil Koperberg would often say to me "You may know about fires but I know about politics".[66]

Koperberg was politically connected with the Carr Government. During another bushfire emergency in Sydney he declared on national television that hazard reduction burning would have made no difference. Rather than embracing responsible land management, the RFS became a top heavy bureaucracy running a paramilitary response organisation around cities with an emphasis on visible buttons and braid, aircraft, fire engines and computers. Koperberg was later endorsed as a Labour candidate, resigned from RFS and was elected to parliament. He served as a Minister for a short time and was embroiled in personal controversy before resigning. He didn't stand for a second term.

Experienced land and fire managers, including rural volunteers, now play second fiddle to green bureaucrats in RFS. Fire hazards continue to increase and megafires occur frequently. (I have heard of the ridiculous situation during fire emergencies in my area, where NPWS officers have insisted that surveys of natural and cultural heritage must precede the clearing of fire trails.) Each summer, television screens are filled with exciting images of firestorms, waterbombers, colourful trucks with flashing lights, and colourful uniforms. Global warming enthusiasts such as Tim Flannery also fill the screens to say "we told you so".

Wilderness enthusiasts believe that Aborigines were 'noble savages' who lived in harmony with nature. Ergo Aborigines did not burn on an industrial scale in disharmony with nature. Fairdinkum scientists recognise that Aborigines initially disorganised and reorganised nature before making burning the mainstay of their economy.

Flannery acknowledges the unassailable evidence of frequent widespread burning by Aborigines.[67] He claims that Aborigines ate the megafauna, allowing vegetation to accumulate and fuel megafires that destroyed most of the rainforests. This supposedly promoted grasses that store little carbon and reflect sunlight, at the expense of trees that store much carbon and absorb more sunlight. Thus Aborigines participated in self-mutilation of Gaia.[68] Flannery thinks that Aborigines were finally forced to use mild fires to protect "what was left of the biodiversity" and thus became "the greatest practical environmentalists on the planet".[69]

But controlled burning is not his favoured approach to sustainable management. Flannery admires megaherbivores because they "supercharged" nutrient cycling as they "shit out copious amounts of dung enriched with ... nutrients".[70] On the other hand, the surviving megafauna apparently aren't so good. Flannery described the elephant as "a great destroyer of forests. Like humans, its original homeland was Africa, and as it spread across the planet 20 million years ago (only Australia escaped colonisation) it too must have affected the carbon cycle". So, although he's keen to introduce carnivorous megafauna like the Komodo dragon into Australia, he doesn't go along with the equally perplexing suggestion of Professor David Bowman that elephants could be introduced to reduce fuels. Flannery allows that we need studies to find out which herbivorous megafauna might work, but "it's self-evident that ill-conceived and poorly monitored programs, such as the proposed trials of cattle grazing in the Alpine National Park, are no substitute".[71]

There were fifteen megafires in southern Australia during the first thirteen years of this century.[72] As Climate Commissioner, Flannery seized on these as examples of the dangers of global warming caused by humans. At about the same time the United Kingdom Meteorological Office quietly announced that there had been no global warming for sixteen years. That didn't deter the Australian Bureau of Meteorology from fixing the broken 'hockey stick' and inventing the *Angry Summer of 2013*. They devised a new temperature statistic that can only be calculated back to 1910.[73] This dispenses with all the extreme records from the Federation Drought and earlier. The sanitised record starts at the coldest period in the last millennium, according to ring widths from Huon pine in Tasmania.[74] Climate hysteria conveniently absolves green academics with a wilderness bent from culpability for the human, environmental and economic disasters that they have visited on Australia through their bad advice to governments on land and fire management.

Endnotes

1 Tench 1793.
2 Jurskis and Underwood 2013.
3 Jurskis 2014.
4 Mitchell 1839.
5 op. cit.
6 op. cit.
7 Ibid., Vol. I, pp. 10-1.
8 Tench 1793, Collins 1798.
9 Collins 1798.
10 Speer *et al.* 1996, Fig. 3.
11 Hurditch and Hurditch 1994.
12 Conroy 1996.
13 Mitchell 1848, Curr 1883.

14 Howitt 1891.

15 Ibid.

16 Jurskis *et al.* 2003, Pyne 2006.

17 Ibid.

18 Adams and Attiwill 2011.

19 Lindenmayer *et al.* 2011.

20 Adams and Attiwill 2011, Ferguson and Cheney 2011, Attiwill *et al.* 2013.

21 Pyne 2006.

22 Adams and Attiwill 2011.

23 Victoria 1939, p. 16.

24 Attiwill *et al.* 2009.

25 Jurskis *et al.* 2006.

26 Merritt 2007.

27 Ibid.

28 Edgar 1969.

29 Costin 1954.

30 Australian Academy of Science 2006.

31 Hudson 1957, McArthur and Free 1968.

32 Merritt 2007.

33 N.P. Cheney pers. comm. 2015.

34 McArthur and Free 1968.

35 Ibid.

36 op. cit.

37 Merritt 2007.

38 op. cit.

39 Anon. 1969.

40 Anon. 1973.

41 Ibid.

42 Leaver and Good 2004.

43 Moriarty 1993.

44 Leaver and Good 2004.

45 Clarke 1860.

46 op. cit.

47 Jurskis *et al.* 2006.

48 Australian Academy of Science 2006.

49 Worboys 2003, Hunter *et al.* 2009.

50 Jurskis *et al.* 2006.

51 Ibid.

52 House of Representatives Select Committee into the Recent Australian Bushfires 2003.

53 op. cit.

54 Adams and Attiwill 2011.

55 op. cit.

56 op. cit.

57 Ellis *et al.* 2004.

58 Adams and Attiwill 2011.

59 Jurskis *et al.* 2003, Burrows 2005, Attiwill and Adams 2008, Boer *et al.* 2009, Adams and Attiwill 2011, Jurskis 2011b, Jurskis *et al.* 2011.

60 Bradstock 2008, Penman *et al.* 2009.

61 Gill 1981.

62 Keith *et al.* 2002, Keith 2012.

63 Jurskis *et al.* 2003, Penman *et al.* 2008, Jurskis 2011b.

64 Bradstock *et al.* 2012.

65 Jurskis and Underwood 2013.

66 N.P. Cheney pers. comm. 2015.

67 Flannery 2012.

68 Flannery 2005, p 17; 2012, p. 42.

69 Flannery 2012, p. 43.

70 Ibid., p. 39.

71 Flannery 2012.

72 Adams and Attiwill 2011, pp. 17-18.

73 Australian Government Bureau of Meteorology 2012, 2013.

74 Cook *et al.* 1992.

8a

Fire research

If you torture the data enough it will confess

Ronald Coase

Some good research, funded by logging for timber and woodchips at Eden, has experimentally confirmed some of the lessons for fire management arising from our ecological history. The Eden Burning Study was initiated and designed by Bob Bridges to examine fire management and silviculture. Doug Binns designed the ecological components of the study. When I took up the position of Officer in Charge of Research at Eden in 1987, I worked with forestry crews to collect baseline data, construct fire trails and apply the experimental treatments. The burning treatments – burning either at two yearly intervals or four yearly intervals or not at all – were applied to areas averaging about thirty hectares. Routine prescriptions normally applied to burning at longer intervals in the forest were applied to both experimental burning treatments in so far as limits of maximum temperature and windspeed, and minimum relative humidity and fuel moisture content were not exceeded.[1]

My involvement in managing the study concluded in 1996 after 130% of the frequent burning area had been cumulatively burnt in three successive fires at two year intervals. At that stage it was apparent that burning had been effective and it was not possible to initiate running fire at two yearly intervals under less than optimal burning conditions.[2] Most of the subsequent experimental burns were conducted when routine burns had been suspended due to suboptimal weather conditions. Over more than a decade when six burns were scheduled and four burns were conducted in the frequent burning area, less than 100% of the area was cumulatively burnt.[3]

These results would confirm to a pragmatic manager, that prescribed burning effectively reduces flammability so that burning at short intervals can and should be done under more severe conditions. They effectively demonstrated how Aborigines could burn safely during summer. However, Dr. David Melick, the new manager of this study, prepared to submit a scientific paper suggesting that burning was a waste of time. Bob Bridges, Forestry Commission Fire Manager Paul de Mar and I pointed out the shortcomings of the paper to Melick as well as his supervisors. He didn't submit the paper to a journal and subsequently resigned from his position. Management of the study came under the control of wildlife ecologist Dr. Rod Kavanagh.

With funds for fire research flowing freely after the 2003 megafires, Kavanagh was able to engage another wildlife ecologist to dust off the old manuscript and further analyse the data from the study. Dr. Trent Penman and colleagues produced a new paper showing that sheltered aspects and positions, as well as recently burnt areas, were less likely to burn under mild conditions than more exposed or less recently burnt areas.[4] This was hardly an advance in knowledge, but it was proof that hazard reduction can work.

Not content with restating these facts, they also applied some fancy statistics. Using data obtained mostly from unsuccessful burns they constructed a computer model predicting that "under operational conditions" ridges are unlikely to burn at intervals less than eleven years and gullies are unlikely to burn at intervals less than twenty years. The outstanding feature of the model was that it was based entirely on fixed environmental parameters such as aspect, slope and upslope catchment. It took no account whatsoever of "operational conditions", because fuel quantity, fuel moisture, soil moisture, air temperature, humidity and windspeed were not included.

Half a century earlier, fire scientists such as Harry Luke, Alan McArthur and George Peet had produced useful guides based on these operational parameters. Penman's contribution was not an advance. It was an example of the dangers inherent in relying on statistics and computer modelling instead of scientific method. If Penman and colleagues had observed fire behaviour and thought about it, they would have realised that the probability of fire burning into a research plot varies not only with environment and time since it was previously burnt, but also hourly, daily and seasonally according to climate and weather. Under true "operational conditions" in dry forests, ridges unburnt for eleven years burn fiercely causing damage to established trees because of high fuel loads. Gullies unburnt for twenty years, especially if they are on sheltered aspects, usually will not burn because there is too much shade, not enough wind and high fuel moisture as a result of understorey development. If it is warm enough, dry enough and windy enough to burn, they generally explode.

Aborigines mostly burnt as soon as fuels were dry enough, so they burnt ridges under mild conditions and gullies under more severe conditions. However, modern managers in southeastern Australia seem to be so tightly constrained by planning requirements, regulations and lack of resources that they are virtually compelled to burn whole blocks under conditions that are unsuitable to achieve the desired results in any part of the block. Things aren't quite so bad in Western Australia where topography is more subdued, but this raises another problem because fires in long-unburnt fuels can travel long distances without being moderated by topography or shelter.

Meanwhile ecologists seem to be looking for reasons not to burn. Penman and colleagues claimed that "Frequent low intensity fire is associated with biodiversity decline"[5] and that high-intensity wildfires are essential to maintain biodiversity.[6] However fire scientists long ago identified the basic physical parameters that should guide prescribed burning to protect our environment, society and economy. Modern scientists including Neil Burrows, Phil Cheney, Jim Gould, Lachie McCaw and Rick Sneeuwjagt have refined our knowledge to the nth degree. But theoretical ecologists are introducing unnecessary complications.

Penman and colleagues claimed that their ineffective burns, conducted under suboptimal conditions, provided refuges for fire-sensitive species.[7] In fact these 'Claytons'[XV] burns allowed fire-sensitive species to

---

XV A non-alcoholic beverage promoted in the 70s as a satisfying alternative to drinkers of alcohol.

escape their refuges and invade naturally open forests, creating a three dimensional continuity of fuels that will eventually ensure that severe fires can penetrate natural fire refuges. For example, in southwestern Australia, after fire was excluded from a National Park for 20 years, all the monadnocks (granite outcrops) that provided natural refuge to rare, fire-sensitive species, were burnt throughout 18,000 hectares by a single high-intensity fire in 2003.[8]

Penman and colleagues thought that "The demonstrated patchiness of fires means that prescribed burning at frequencies which exceed thresholds derived from considerations of life history attributes of fire-sensitive species do not necessarily imply a high risk of local extinction, as is often assumed".[9] But it's not the very fine scale patchiness of ineffective burns that protects fire-sensitive species. To be effective, burning must reduce the fuel right through the matrix, leaving only the physically protected sites such as swamps, rock outcrops and deep dark gullies unburnt.

Penman further claimed that "frequencies that are high relative to such thresholds do reduce the extent of potential refuges and increase the risk of local extinction".[10] This conclusion is wrong because frequent mild fires do not impact on the real refuges that are physically protected from fire. The extent of these refuges is fixed by topography, soil, hydrology and climate. Frequent, **effective** fires actually reinforce the integrity of the refuges by ensuring that fuels in the surrounding landscape are low and discontinuous. This is why the ecologists were unable to achieve running fires under suboptimal conditions for burning. It is the reason why Melick thought that frequent burning was a waste of time. It is the reason why Penman thought that ridges wouldn't burn at intervals less than 11 years and gullies couldn't be burnt at less than 20 year intervals.

A second paper from the Eden study should have emphasised that frequent burning promotes biodiversity. Penman and colleagues discovered that species richness of plants increased with the number of fires and decreased with time since fire.[11] Unsatisfied with these results, they analyzed species richness over time and found that it declined across all treatments.[12] Species richness declined in the burning treatments over time because the intended treatments were not achieved and few sample plots were burnt each time. Inappropriate statistical models allowed the fully sampled time factor to overwhelm the grossly undersampled fire factor. Penman and colleagues claimed that "there was a general decline in plant species richness at both coupe and plot scales which occurred independently of imposed management regimes".[13] This opinion was published in a reputable scientific journal together with the factual statement that burning increased species richness, even though it seems that one statement contradicted the other.

Worse still, was the following illogical speculation in the abstract of the paper: "This [alleged general decline in plant species richness] is thought to be related to a natural succession following wildfire, and may be due to the absence of high-intensity fire in the study area since 1973, or to an effect related to season of burning". In fact, the 1973 wildfire burnt at a low intensity through the whole study area,[14] and the experimental treatments, when actually achieved, increased species richness despite the long absence of high-intensity fire. It is neither the frequency nor season of fire that counts. It is the physical impact. Mild fire that runs freely and kills woody seedlings but not mature trees and shrubs, maintains biodiversity. High-intensity fires create dense stands of a few shrub species that quickly choke out small shrubs, herbs and grasses.[15]

Penman and colleagues analysed another dataset to 'confirm' their speculations about the benefits of wildfires. This paper began with a highly provocative statement that "Wildfire has shaped historic … vegetation assemblages in Australia". The analyses clearly showed that increasing competition from developing shrubs caused a decline in diversity of groundcover plants over thirty three years after wildfire.[16] Prescribed burning wasn't tested in this study. Nevertheless the results were entirely consistent with the results from the burning study showing that species richness increased with burning and declined with time since burning. Penman's conclusion that wildfires are essential to maintain species richness was wrong. Wildfire has shaped vegetation only since Europeans disrupted the natural regime that applied for 40,000 years.

Species richness peaks about four years after fire in dry eucalypt forests.[17] Following this peak, increasing densities and stature of a few common shrubs start to shade out a wide range of smaller plants. Other studies in these forests suggest that intervals of about five years between mild fires maintain the nutrient cycling processes that keep established eucalypts healthy.[18] The Eden Burning Study also showed that eucalypts declined in the absence of frequent mild fire, initiating plagues of mistletoes in ridgetop stands, and plagues of psyllids and bellbirds in gully stands.[19]

Matthias Boer and colleagues analysed the results of half a century of routine burning in the eucalypt forests of southwestern Australia. They showed that prescribed burning effectively reduced fire hazards for about six years.[20] Obviously, increasing shrub development, increasing fire hazard, declining species richness and declining eucalypt health are inextricably linked. Another study by Janet Farr and colleagues in southwestern Australia showed that outbreaks of jarrah leaf miner did not occur in forests that had been deliberately burnt within the previous three years.[21] All these studies showed that frequent mild burning, at intervals of about three to six years, maintains biodiversity, health, resilience and fire safety in dry forests. David Ward's seminal studies of fire markers in grasstrees pointed to fire intervals of about three years under Aboriginal management of jarrah forests.[22]

Ward developed a technique of grinding back the charred leaf bases on stems of grasstrees to reveal alternating annual growth bands and black bands left by fires. Inevitably, Ward ran into trouble with proponents of the circular argument around obligate seeding shrubs, chiefly Neal Enright. Ward's method gives a fire history specific to each individual grasstree examined. We know that Aborigines burnt in a mosaic pattern, and sometimes burnt grasstrees individually. Enright compared Ward's grasstree histories against fire histories derived from satellite images. He claimed that differences between the coarse satellite record and the precise grasstree record indicated "anomalies" in the grasstree records.[23] That's a bit like using a compass and pacing traverse to challenge the accuracy of a survey conducted with a laser theodolite.

When I visited Western Australia, Ward pointed out to me that the few remaining old grasstrees in many conservation reserves often have shaggy skirts of unburnt dead leaves hiding rotting trunks that make it difficult to apply his technique. The situation was quite different under Aboriginal management. When Charles Darwin described a shrubland of the type that Enright claims is threatened by frequent burning, he noted that the grasstrees carried "merely a tuft of very coarse grass-like leaves". They had no skirts, and the "general bright green colour of the brushwood and other plants, viewed from a distance"[24] was clearly indicative of a stand that had a flush of new shoots after being lightly scorched by a fire of low to moderate intensity.

Ian Abbott compiled historical data showing that the fire regime described by explorers and early settlers[25] matched the prehistoric records from grasstrees. In eastern Australia estimates of pre-European fire regimes derived from historical accounts[26] are consistent with Ward's and Abbott's data from the south-west. Prehistory, history, fire science and ecology all show that mild fires at intervals of less than five years maintain biodiversity, ecosystem health, resilience and fire safety in dry eucalypt forests.

This body of science has repeatedly been challenged by studies of charcoal accumulation or dendrochronology of fire-scarred trees. But fire-scarred trees or sedimentary charcoal do not provide a satisfactory record of fire regimes because they do not faithfully record the occurrence of low-intensity fires in open grassy ecosystems.[27] A seminal study of fire scars in snow gum stands by John Banks compounded this problem because he combined records from ancient woodland trees at one site with dense young forest trees at other sites, to 'demonstrate' an increase in frequency of fires after European settlement.[28] All that the data really showed was that more intense fires in heavier fuels more easily scarred thinner barked younger trees in dense stands than did mild fires in naturally grassy snow gum woodlands.

Prescribed burning experiments in Australia typically show that burning either favours biodiversity or does not affect it.[29] Imlay mallee (see Chapter 6) and the orchid *Prasophyllum correctum*[30] are critically endangered because of long-term fire suppression. The rare and endangered Hastings River mouse and eastern brown treecreeper have disappeared from new national parks that were protected from grazing and burning.[31] Habitat features that support the superb parrot such as scattered ancient red gums in open grassy woodlands are being lost to woody thickening and chronic tree decline in unburnt remnant ecosystems.[32] Endangered broad-headed snakes have been lost from national parks around Sydney and are declining in national parks further south because lack of burning has allowed scrub to choke out their basking habitat.[33] Even common species such as grasstrees and Tasmanian devils are now being lost because we no longer maintain their open habitats using mild fire (Chapter 6).

Surprisingly, reports of the Kapalga experiment in the Northern Territory painted a different picture. Alan Andersen and colleagues reported a "serious conservation concern" that riparian vegetation and small mammals were not resilient to fire, irrespective of intensity.[34] However Andersen's team apparently didn't understand fire behaviour and consequently failed to properly test fire intensity. Their "early", "late" and "progressive" fires were all lit using a single line head fire, two to five kilometres along the windward side of each block[35] so that the effect of seasonal curing was overwhelmed by the inappropriate ignition pattern.

Fairdinkum fire scientists such as Phil Cheney and Jim Gould of CSIRO, and Lachie McCaw of Department of Parks and Wildlife in Western Australia established that fire spreading from a line more than 100 metres long across the wind would reach its full potential rate of spread within 100 metres. They factored this into the design of their *Project Vesta* wildfire behaviour project[36] to ensure consistent and credible outcomes. It's not surprising that Andersen's Kapalga burns impacted some small mammals and riparian zones, because every fire consumed virtually the entire treatment area at the maximum possible intensity under the prevailing physical conditions.

Pragmatic fire managers would have used a different lighting pattern such as multiple spot fires lit progressively into the wind to moderate the intensity of the burns. Instead of reexamining their ignition techniques and timing, Andersen's team proposed that it was necessary to increase the proportion of landscape remaining long-unburnt by burning less each year and by not burning long-unburnt areas. They failed to heed the lessons from ecological history and failed fire management in southeastern Australia.

Thankfully one of Andersen's colleagues, Jeremy Russell-Smith, 'got the message' and started a project in 1995 to reintroduce the Aboriginal firestick to nearly three million hectares of West Arnhem Land by progressive burning as fuels gradually dry across the landscape. Over seven years from 2005 they not only enhanced biodiversity and the local Aboriginal socioeconomy but they also reduced emissions of greenhouse gases by a third compared to the ten year baseline period of conflagrations late in the dry season.[37]

Martu people of the Western Deserts have been quietly working to restore traditional burning and native animals in their vast country since they started to return from mission settlements in the late-twentieth century. After they left their country to make way for testing of British 'intercontinental' ballistic missiles, megafires ignited by lightning replaced patch burning by people, and 18 different animals were lost from their country. Neil Burrows sent me a Youtube video about this project[38] late in 2014 when wildfires started by lightning were successfully contained within about 5,000 hectares as a consequence of traditional burning during the previous decade. Neil told me that the wildfires had the potential to cover at least ten times that area and destroy the work that had been done so far to reintroduce native animals lost during the mission period.

This video is enlightening for anyone interested in firestick ecology. Martu people have a story of "relations between fire and rain – forces that are mates to each other. Smoke can bring rain; rain feeds grasses; dry grasses and shifting winds feed fire".[39] To me it's a statement of the 'circle of life', and man with his firestick is central to it.

Endnotes

1 Binns and Bridges 2003, Bridges 2005.

2 Jurskis *et al.* 2003.

3 Penman *et al.* 2007.

4 Ibid.

5 Penman *et al.* 2008.

6 Penman *et al.* 2008, 2009.

7 Penman *et al.* 2007.

8 Burrows 2005.

9 Penman *et al.* 2007.

10 Ibid.

11 Penman *et al.* 2008.

12 Ibid.

13 Ibid.

14 Binns and Bridges 2003.

15 Penman *et al.* 2009, Fig. 1.

16 Ibid.

17 Penman *et al.* 2008, 2009.

18 Turner *et al.* 2008.

19 Jurskis *et al.* 2005, Jurskis 2011b.

20 Boer *et al.* 2009.

21 Farr *et al.* 2004.

22 Ward *et al.* 2001.

23 Enright *et al.* 2005.

24 Darwin 1845.

25 Abbott 2003.

26 Evans 1814, Darwin 1845, Mitchell 1848, Curr 1883, Howitt 1891.

27 Burrows *et al.* 1995, Hassell and Dodson 2003.

28 Banks 1989.

29 Wittkuhn *et al.* 2011.

30 Coates *et al.* 2006.

31 Tasker and Dickman 2004, Ford *et al.* 2009.

32 Jurskis 2005, 2009, 2011a; Manning *et al.* 2006.

33 Pringle *et al.* 2009.

34 Andersen *et al.* 2005.

35 Ibid.

36 Gould *et al.* 2007.

37 Russell Smith *et al.* 2013.

38 Walsh 2013.

39 Ibid.

9

THE ALLEGED EVILS OF GRAZING

All of Rubin's cards were marked in advance
The trial was a pig-circus he never had a chance
Bob Dylan – *Hurricane*

Grazing was wrongly convicted, by association, of causing 'New England Dieback' when prosecuted by naturalist A. Norton during the Proceedings of the Royal Society of Queensland in 1886.[1] More than a hundred and twenty years later it was wrongly convicted of a similar crime, in Tasmania's Midlands, for similar reasons. As we saw in Chapter 6, rural tree decline is an example of a global phenomenon of species declines and pest irruptions caused by accumulation of nitrogen in the soil. It is associated with grazing where grazing relies on improved pastures. Sowing and fertilising leguminous, nitrogen-fixing pastures rapidly alters the chemistry and physics of soils making them unsuitable for species whose natural habitat was maintained by the firestick. In the recent Tasmanian study, Dr. Dugald Close and colleagues acknowledged that their intensively-grazed treatments "contained exotic improved pasture maintained by fertiliser applications"[2] but they unaccountably attributed differences between treatments to intensity of grazing.

It has long been the wont of Australian ecologists to attribute most environmental problems in rural lands to the evils of grazing. The list includes soil compaction, soil erosion, loss of biodiversity, loss of vegetation 'structure' and chronic eucalypt decline. Having seen compaction and erosion of soil from laneways traversed four times daily by dairy herds, and less spectacular damage around some watering points, I am aware that stock can cause soil damage. Close's study attributed soil compaction to intensive grazing. Here again, it was the cultivation employed in sowing pastures that caused the problem. The soils under native pastures weren't compacted.

Gammage also 'pointed the finger' at soil compaction by grazing stock:

> Change 1. Since 1788 compacted soil and speeding water have constricted water sources and the foods they nourished. The earth has changed. Topsoil blows away, hills slip, gullies scour, silt chokes, salt spreads, soil compacts. The last is least noticed. Much of 1788's soil was soft enough to push a finger into – 'naturally soft', Thomas Mitchell called it.[3]

But much of our soil is still naturally soft because it is sandy. Mitchell was on a sandplain near the Darling river when he made the remark quoted by Gammage. Much of the soil that Mitchell described was not soft. After he crossed the sandplain he firstly found hard soil, and then firm soil:

> At length, we gained some gentle rises, at the base of which, the soil was a clay, so tenacious as to have hollows in its surface, which during wet seasons, had evidently retained water for a considerable time. ... The round knolls consist of a red earth, which is different from the soil of the plains ; its basis appearing to be iron-stone. We encamped on good firm ground.[4]

Curr attributed "the baked, calcined, indurated condition of the ground so common to many parts of the continent, the remarkable absence of mould which should have resulted from the accumulation of decayed vegetation, the comparative unproductiveness of our soils"[5] to Aboriginal burning. Flinders was impressed by the quality of the "vegetable soil" on uninhabited Flinders Island.[6] Mitchell noted the absence of "vegetable soil" due to Aboriginal burning on Sydney sandstones.[7] McIntosh's study of soil formation processes in Tasmania and New Zealand showed that Aboriginal burning produced relatively infertile, poorly structured soils well suited to plants adapted to aridity.[8]

Some soils became harder after European settlement because topsoil was lost after overgrazing by feral rabbits and domestic stock during droughts. Gammage attributed to compaction, these changes described by a grazier at Cobar in 1901:

> a gradual deterioration of the country caused by stock … has transformed the land from its original soft, spongy, absorbent nature to a hard clayey smooth surface (more specially on the ridges), which instead of absorbing the rain runs off it in a sheet as fast as it falls.[9]

The 1901 Royal Commission into the Western Division of New South Wales recognised that the hard red ridges around Cobar were caused by drought, rabbits, overstocking and sand storms which blew the topsoil away.[10]

Domestic stock in combination with feral rabbits and droughts caused severe and extensive erosion and dust storms across southern Australia between the late-nineteenth and the mid-twentieth centuries. Since then most of the environmental problems in rural lands stem from either pasture improvement or from not grazing and/or burning.[11] Over three decades I have seen extensive areas of healthy native vegetation turn to scrub with dying trees after grazing was excluded. I have also been distressed to witness a huge waste of taxpayers' money on ecological restoration schemes where long corridors are fenced and planted to native trees and shrubs, creating imaginary native ecosystems that cannot function in the absence of grazing or its more natural ecological analog of mild fire.

The historical context of Dr. Alec Costin's campaign to exclude grazing and burning from the 'High Country' indicates that it did not start from objective scientific observations, as evident in the following extracts from an interview with David Salt published by the Australian Academy of Science:[12]

> What was the aim of the Soil Conservation Service? Was it to help farmers?
>
> It was, but actually it was the brainchild of Sam Clayton, the first Commissioner, who was a senior agronomist at that time with the New South Wales Department of Agriculture. In his opinion there

was enough land degradation to justify a big soil conservation component in extension work of the Ag Department, but the department was not interested.

Then Clayton got a scholarship to the United States during the time following the dustbowl/Depression years where conditions were described in The Grapes of Wrath by Steinbeck. He came back all fired up that something had to be done. Again the department was not interested, but because of Clayton's friendship with the Minister for Mines the legislation was enacted in 1938 and the Soil Conservation Service did come into existence – at first with a very small group of people who had a lot of miles on the clock. And from the time I was given the cadetship, at the start of my agricultural course in Sydney University, I was already a fairly senior public servant – not that being senior meant anything to me.

You studied for a Bachelor of Science in agriculture at the University of Sydney. For your honours year you studied the ecology of the Australian Alps, didn't you?

Yes. I did quite well during my four years' pass course, and it was suggested by Noel Beadle, a lecturer in botany who was already a colleague of mine, and by Gordon Hallsworth, a lecturer in soils in the Agriculture Faculty, that I should consider an honours year in agriculture or botany. I really didn't have much idea of what I could or should do, but it had to be in the general framework of the course subjects in the Faculty of Agriculture. Then the Botany Department pointed out that I could have a go at the Australian Alps, where they said very little work had been done and where in particular virtually nothing at all was known about the soils. That suggestion got the green light from Sam Clayton, who was already deeply concerned about the condition of high mountain catchments in New South Wales and had participated in quite extensive field inspections of Kosciuszko which led to legislation enacted in 1944, setting up the State Park there. And although there was a recreational basis for setting up the park, the primary reason for doing it was catchment protection, so that early legislation was soon followed by the first moves to reform snow lease grazing in the mountains. Clayton, then, was more than happy that I should start an honours year in an area that he himself was concerned with.

Had you had any experience of working in the High Country?

None whatsoever. I'd only seen snow once, I think at Katoomba, and I had no idea of what was ahead. But I find that the more you get involved in a subject, the more it sucks you into it, and that was especially so from the soils and vegetation points of view. The country was just magnificent, and because almost everything that one looked at was new – there was so little known, and not much help in the literature – this was a real voyage of discovery. Almost everywhere you looked or anything you did, it was a challenge to follow up and find out a bit more.

Cooma would have been a fairly small town at that time, in the late 1940s.

That's correct. It was a small, very stable, very inbred, very interesting place to be.

And I think you had the first bulldozer in the area.

Yes. This was just post-war and there was very little earthmoving equipment in Australia, but the Soil Conservation Service managed to latch on to a couple of bulldozers. One of them was sent down from Goulburn, where there was a main regional office of the Service, to start up a bit of work around Cooma. In particular, there was some very badly eroded land at Bredbo, and what Sam Clayton was very keen to do – and he did it well – was to set up demonstrations in soil conservation on highways et cetera where the public couldn't help noticing them. So a little bit of my work was involved with that, but essentially I was meant to continue the vegetation, soil and land use surveys.

I would summarise the historical context as follows: The head of the SCS had seen huge erosion problems created by farmers in Kansas, and similar problems created by farmers (and rabbit plagues) at Bredbo on a smaller scale. He organised demonstrations of remedial earthworks at Bredbo. At the same time, having responsibility for advising management of catchments in the Hume-Snowy hydroelectric, irrigation and flood mitigation scheme, he inspected the catchments soon after the 1939 megafire and wrongly assumed that erosion caused by mining and wildfires, had been caused by seasonal grazing. There was no scientific information to guide catchment management in this area. He appointed a young graduate to gather some information. This person did not appreciate the knowledge and experience of the local community, and had no relevant scientific literature to guide him, so he set out to confirm his boss's assumption that grazing and burning caused erosion, despite his observation that the country was "just magnificent".

While working for the SCS between 1947 and 1950, Costin established some plots to monitor the soil and vegetation conditions in the High Country and wrote an honours thesis. When Clayton recalled him to Sydney, he resigned and obtained scholarships which allowed him to turn his thesis into a book and to look into alpine ecology in Europe. Later on Costin was asked "How did the high country overseas compare with Australia's high country?".[13] His response shows that looking at a totally different environment with a completely different history somehow served to reinforce his prejudices against grazing in the Australian Alps:

> That's a big question. As far as ecology is concerned, the overseas mountains were generally much steeper, much younger as the result of more recent uplifts, and also much younger as regards the development of soils et cetera. Mostly the environments that the overseas mountains presented you with had developed since the Ice Age, whereas most of our mountains survived glaciation. …
>
> As far as land use was concerned, almost all of those mountains had been used for transhuman grazing for centuries and so many of them had developed a quasi-equilibrium, but almost without exception they were obviously grossly disturbed. Those in the Australian Alps showed very much less disturbance. For instance, most of the British mountains used to be forested but are now covered with heathland. Most of the European Alps used to be forested, and one reason they are so prone to snow slides and avalanches is that most of the timber is gone. So, from the point of view of land use and its effects, the Australian Alps were substantially different from most of those I worked in overseas.
>
> Seeing these things overseas reassured me a little bit. In my reports

up till that time, especially in the Monaro ecosystems book, I had trodden very lightly indeed, being very cautious about cause and effect and about the relation between high country grazing and catchment deterioration. I now realised I had been soft-pedalling, and this had a big influence on what I did later – not that I was trying to prove anything, but future work needed to recognise that the controlling factors in what had to be done were human as well as natural.

In the "Monaro ecosystems book",[14] Costin noted that "the intensity of rabbit grazing greatly exceeds that of domestic livestock" in much of the tablelands. But he was concerned about "the prevention of tree regeneration", and didn't mention the massive contribution of rabbit burrows to erosion gullies. A Victorian High Country cattleman and Shire President by the name of Frank Blair recounted that rabbit plagues and the Federation Drought followed by wet seasons initiated scrub development which was exacerbated by the Forests Commission's prohibition of burning. This set the scene for severe fires in 1925 and 1926 that destroyed alpine bogs and caused massive erosion.[15]

When he returned to Australia, Costin was employed in Victoria's High Country. But during that time he:

> revisited some of my original sites on the Main Range, where right from the beginning I'd been very careful to keep permanent records – initially photographic records and later transect and quadrat measurements that were repeated over a long period of years. These sites had been taken out of grazing almost as soon as the Kosciuszko legislation was passed, so they were already protected from the main disturbing agent. But I found that in the few years since I first saw them the amount of degradation had increased incredibly, simply reflecting the fact that if environments are near the edge of stability and you just tip them over the edge, their condition will continue to go downhill.[16]

An objective assessment of the increasing degradation would suggest that wildfires had been the main problem, and this problem got bigger after grazing was eliminated. The 1951/52 wildfires compounded erosion problems from wildfires in 1926 and 1939. Furthermore, survey and construction work on the Snowy Scheme had initiated spectacular localised erosion by the time Costin revisited the Snowys. It was reported that "over 280,000 cubic yards of silt … had washed down the Tumut River from the Cabramurra town site".[17] (Cabramurra is the highest settlement in Australia at 1481 metres.) Noel Gough's book *Mud Sweat and Snow* includes several anecdotes and photos of horrible boggings and extractions of vehicles and heavy machinery.[18] But Costin wrote a report about the 'increasing degradation' and "a copy went to the Snowy Mountains Authority, which received it quite well despite the inclusion of some criticisms of the Authority's own works at the time".[19]

Had he followed the scientific method, Costin might have modified his hypothesis that grazing was the big problem rather than calling on 'tipping points' to defend his prejudices. But this is an illustration of the quality of "the absolutely overwhelming evidence that rangeland grazing … has no part in our higher mountain catchments".[20] Snowy Mountains Man, Tom Taylor, guided Sam Clayton's and Premier William McKell's inspections of the High Country in the early 1940s. On Clayton's recommendation he also guided Costin during his work for SCS whilst compiling his thesis. When Costin's book was published in 1954, its conclusion was that grazing had to be curtailed and fire had to be suppressed to save the mountains from erosion.[21] The snow-lessees knew that grazing and burning snow grass actually prevented soil erosion. They pointed out to the academics and bureaucrats that "The persisting erosion in areas that had not carried stock for years, contrasted sharply with a denser grass cover elsewhere".[22] Taylor told Costin "You have done a lot of work young Alec, but you've taken a wrong turn".[23]

On the other hand, Chief of Snowy Mountains Hydroelectric Authority (SMHEA), Bill (soon to be Sir William) Hudson, was very happy to join a developing anti-grazing coalition, possibly to divert attention away from severe erosion caused by survey and construction work.[24] He collaborated with Murray Basin irrigators, the newly formed Australian Academy of Science and anyone else opposed to grazing in alpine catchments. University House in Canberra became an old boys' club for meetings of academics and bureaucrats with SMHEA. Meanwhile Clayton mostly worked behind the scenes, with the close confidence of the Premier of NSW to support the anti-grazing coalition who portrayed alpine graziers as greedy and unpatriotic.[25] In 1954 Costin was appointed as CSIRO's specialist in High Country ecology, allowing him to compile further 'evidence' of the evils of grazing and burning. SMHEA contributed to his salary.[26]

In 1957, the Australian Academy of Science carried out inspections of the High Country, and in less than three months produced a report endorsing the claims of the anti-grazing coalition.[27] Hudson provided transport for the inspections, and chaperones in the form of SMHEA soil conservation officers. Costin was a member of the Academy committee, and the driving force behind the report. The Chairman of the committee, J.S. Turner, was Professor of Botany at Melbourne University. He later contributed a chapter to Marshall's assault on pastoralists, farmers, foresters etc. – *The Great Extermination*. Turner wrote to more than twenty organisations having views opposed to grazing and compiled a lengthy chapter in the report legitimising their collective view. In his historical account of this "first major political clash in Australia over the environment", John Merritt wrote that "Anyone reading this chapter could only have concluded that in the opinion of well-respected authorities grazing was endangering the national interest". Grazing above 1350 metres was banned in 1958.[28]

At this time, opposition MP Tom Lewis assured lessees that he was in favour of controlled grazing. After elections in 1965, new Minister for Lands, T.L. Lewis launched a master plan for management of Kosciusko State Park that called for elimination of grazing. In 1967 Lewis introduced a bill to establish the National Parks and Wildlife Service with responsibility for all parks in New South Wales including the soon to be rebadged Kosciusko State Park. In his second reading speech, he referred to consultations with National Parks Association, National Trust, Wildlife Preservation Society, Sydney Bushwalkers and Dunphy's defunct National Parks and Primitive Areas Council.[29] He also referred to Marshall's recently published book *The Great Extermination*, but there was no mention of consultations with grazing interests because there weren't any.

In the run up to elections in 1968, lessees extracted promises from both sides of Parliament to conduct independent inquiries into High Country grazing. After the elections, Lewis appointed veterinary scientist Dr. Grahame Edgar to conduct an inquiry. Edgar took almost twelve months to compile a report restating the views of Costin and the Academy that grazing should be prohibited in Kosciusko National Park. Historian John Merritt suggested that a week would have been ample time to summarise the Academy's report and any of Costin's articles. There appears to be little doubt that the result of the inquiry was preordained. Merritt's history of the twenty-five year battle makes it easy to understand why High Country graziers and pragmatic land managers came to regard the anti-grazing and anti-burning ecologists as "arrogant bastards" and "dangerous fanatics".[30]

Merrit's book published in 2007, provided the first detailed and balanced account of the battle but unaccountably neglected to report the ongoing socioeconomic and environmental outcomes. The horrific consequences of disastrous fires in 1978, 1983, 1988 and 2003 left no doubt about the spurious scientific arguments against grazing and burning. I don't think anyone has been brave enough to make public an estimate of total soil loss in the High Country immediately after the 2003 megafire. But hydrology professor Jacky Croke estimated that up to 200 tonnes of soil per hectare eroded from some catchment slopes.[31] This severely reduced the water quality and capacity of Cotter Dam. Firstly a new filtration plant, and then a new wall had to be constructed to maintain Canberra's water supply.

A wilderness mentality is so strong in many environmentalists that they will damage their own private lands by excluding grazing and fire, and not see the problems they have created. This is evident in the following extracts from a recent Feature Article in *Ecological Management and Restoration*:

In the 1970s, within the forests of the broader Toonumbar area

west of Lismore in north-eastern NSW, there was little Lantana
(*Lantana camara*). On the ridges, the drier eucalypt forests had grassy
understoreys; further downslope, understoreys were shrubby;
and in the wetter gullies, rainforests or rainforest elements were
common. The native honeyeater Bell Miner (*Manorina melanophrys*)
was present but not widespread. By the 1990s, due mainly to
previous logging disturbance, much of the forest understorey had
been replaced by Lantana (50–90% cover), and around the same
time, Bell Miners became common throughout most of the forest.
The hills rang with their incessant and penetrating calls from dawn
until dusk; other birds were scarce; tree canopies were shrinking,
and some trees were dead…

… During the 1960s and 1970s, the previous owners selectively
logged much of the forested area and low-intensity grazing by
cattle followed. When we purchased the property in 1980, cattle
were excluded from the forests and the forests were left to
regrow. However, their natural recovery was hampered by Lantana
invasion. By the 1990s, the Bell Miner had established large and
persistent colonies on the property and exploited for food the
increasing populations of Glycaspis psyllids, which were feeding
on the eucalypts.[32]

This article was written by a couple who clearly initiated chronic
eucalypt decline, lantana invasion and plagues of psyllids and bellbirds
on their hobby farm by excluding grazing and fire. With their "bush
regeneration contractor", they wasted a lot of money on the secondary
problem of lantana invasion without addressing the fundamental issue
of changes in soils and nutrient cycling caused by exclusion of grazing
and/or burning. In the 1970s the forests were logged, grazed, frequently
burnt and healthy. The 'tree changers' excluded grazing in the 1980s,
and by the 1990s the trees were declining, and lantana, psyllids and
bellbirds were proliferating. Had they substituted frequent mild burning
for grazing they wouldn't have had a problem. The adjoining forests of
the "broader Toonumbar area" contain some examples of healthy, grassy
forests that are logged, grazed and burnt and many unhealthy, lantana-
infested forests that are neither grazed nor burnt.

I'm astounded that only a tiny minority in academia, bureaucracy and 'the media' can see the widespread and striking contrasts in the health and resilience of native vegetation that is grazed compared to native vegetation that is supposedly protected in ungrazed and unburnt reserves. I don't like to look at 'outdoorsy' shows on television anymore because they almost invariably contain scenes with backgrounds of scrub and dying trees. Maybe the epidemic of voluntary blindness arises from the availability under the Whitlam government of free tertiary education which encouraged students looking for a cause rather than a career. Much of the current crop of professional greens was sown at that time. These people confuse homogenous biomass with biodiversity, and horizontally and vertically continuous fuel with 'ecosystem structure'.

The narrow-minded attitude of conservation bureaucracies to grazing is evident in this extract from the National Forest Policy Statement: "Grazing in public native forests can have a significant impact on and severe implications for forest ecosystems. In particular, grazing in National Parks or reservation areas is usually inconsistent with protecting conservation values".[33] Some of our leading green academics went even further when 125 "concerned scientists" opposed small trials of traditional seasonal grazing in the Victorian Alps, and wrote to the environment minister that: "the negative ecological impact of cattle on Australian native ecosystems is well documented".[34] In my opinion, an objective scientist would not put their name to such a breathtakingly sweeping statement, but it was signed by an impressive array of Professors, Doctors and lesser mortals. If I were seeking scientific advice on the subject of grazing, I would employ this list as a rogues' gallery, but Federal Minister for the Environment, Tony Burke, intervened to prevent the trials.

The "well documented" adverse impact of cattle grazing is based almost entirely on a circular argument. Scientists have ignored our ecological history. They assume grazing is bad for the environment and compare grazed woodlands against 'benchmarks' derived from ungrazed and unburnt vegetation that in no way resembles pre-European woodlands. Grazed areas are quite different to these unmanaged scrubs, ergo grazing is bad. But beneficial impacts of grazing in the landscape are quite obvious to anyone with an open mind.

Whether the overall impact is positive or negative, needs to be assessed against the type of ecosystem, the intensity of grazing and a realistic picture of the pre-European environment. One 'Concerned Scientist', Dr. David Keith, claimed that grazing and burning depleted shrub understoreys on New South Wales' northern tablelands escarpment[35] because he imagined that scrub-infested reserves represented natural conditions. Another study in the same area demonstrated that the rare and endangered Hastings River mouse lived only in grazed and burnt country[36] where its naturally open and grassy habitat had been maintained, and not in the scrubbed-up reserves that were overrun by bush rats.

The New South Wales Office of Environment and Heritage banned grazing and burning from potential habit of Hastings River mice on State forest. But when they bought a property in south-western NSW to conserve the plains wanderer they showed a different attitude:

> The moderate grazing levels on this former merino stud property have afforded the long-term survival of the plains-wanderer by simulating the ideal grassland habitat no longer present in other areas. To ensure that the plains-wanderer habitat on Oolambeyan is maintained and enhanced, ecologically sensitive sheep grazing is being conducted by the NPWS. This will ensure that the plains-wanderer habitat at Oolambeyan is kept in good condition even during droughts when stock are removed.[37]

The only explanation that springs to my mind, for the different attitudes, is the high level of public scrutiny attached to private land purchase for specific conservation goals. In public forests, there are no previous owners to keep an eye on the new managers, and most people wrongly assume that conservation bureaucrats know what they're doing. Furthermore, greens have disproportionate political influence and constantly push the wilderness philosophy. So minimising human economic activity gets higher priority than conserving threatened species.

The concept of "ecologically sensitive sheep grazing" provides a nice counterpoint to the "well documented … negative ecological impact of cattle". This might help to explain why Victoria legislated in 2015 to ban only cattle grazing from alpine and river red gum national parks. But positive impacts of cattle grazing are quite obvious in both alpine and river red gum ecosystems.

Some increasingly rare woodlands of river red gum that have been grazed since pastoralists disrupted Aboriginal burning 150 years ago have a natural structure with a canopy of healthy, ancient, spreading trees. The NSW Natural Resources Commission assessment of river red gum[38] was predictably biased against grazing, but the report used an image of one of these ancient woodlands to exemplify a "stand of healthy River red gum forest". I had used a different image of this same stand in a scientific paper[39] to characterise a healthy woodland (Fig. 7). The image in the assessment report shows a patch of saplings amongst the ancient trees. Ecologists typically regard the presence of saplings and seedlings as evidence of a healthy system. But these saplings are in a

Fig. 7 Ancient, grazed river red gum woodland at the height of the Millenium Drought, 2009. (Courtesy of NSW Department of Primary Industries)

large canopy gap created by treefall. The remains of the butt of a fallen tree are visible in the picture. Saplings are generally absent from healthy woodlands except where a canopy opening has created an opportunity for regeneration. They are abundant in ungrazed and unburnt woodlands because the old trees are sick and their canopies are declining.[40]

The healthy red gum woodland is in a State forest that has been grazed under lease to the owner of adjoining private land, and is popular with campers and fishermen who use fallen timber for firewood – another activity frowned upon by green academics and bureaucrats. The ancient stand is healthy because grazing, camping and firewood collection are ecologically analogous to Aboriginal management.[41] By contrast, protected stands typically contain very few ancient trees, all on their last legs, amongst a forest of chronically declining post-European trees engulfed by cherry scrub. Cherry, a parasite on red gum roots, is one of the arbivores that flourishes when canopies decline in the absence of grazing or frequent burning.

Healthy, grazed, ancient woodlands exemplify the purportedly "well documented … negative ecological impact" of grazing. For example, a report by Roberston and Rowling claimed that grazing has altered the structure and function of river red gum stands because:

> Seedlings and saplings of the dominant *Eucalyptus* tree species were up to three orders of magnitude more abundant in areas with no stock access, and the biomass of groundcover plants was an order of magnitude greater in areas with no stock.[42]

Similarly Jansen and Robertson found that ungrazed stands were in 'good condition' because they were full of shrubs, leaf litter, fallen timber and red gum seedlings – a sure sign that they were in fact chronically declining.[43] As usual, the academics imagined a pre-European landscape vastly different to that described by explorers and naturalists such as Mitchell and Curr.

'Concerned Scientist' Ian Lunt and colleagues[44] referred to these fatally flawed studies to support their claim that "The negative impacts of stock grazing on riparian *E. camaldulensis* forests are well documented". But they suggested that "the potential for ecosystem recovery following the removal of grazing stock is poorly known". So they conducted a trial where grazing was excluded from some "degraded, annual-dominated understorey vegetation in riparian forests dominated by *E. camaldulensis* in the Gulpa Island State Forest". They monitored grazed and ungrazed areas for twelve years and found that grazing reduced fuel (total plant cover), but had no impact on the number of native or exotic plant species. Lunt and colleagues claimed that this is what you would expect because the site was highly degraded and unproductive.[45] However they provided no information about the natural vegetation prior to the introduction of grazing, and no evidence of degradation by grazing.

The natural vegetation in this area was mostly swamps, grasslands and reedbeds.[46] The red gum forest owes its existence to disruption of Aboriginal burning after 1842 and reduced flooding since the early twentieth century.[47] It lacks species richness because red gum dominates the site and a few robust species dominate the ground cover.[48] It is dominated by annuals because regular flooding kills seedlings[49] and delivers new seeds. A recent study of changes in the understorey of the Barmah Forest showed that that reduced flooding increased the richness of both native and exotic species.[50] Lunt's data simply confirmed that the tree canopy, flooding and seasonal conditions shaped the ground cover, and the only real impact of grazing was to reduce fuel.[51] In a major review of potential grazing impacts, Lunt and colleagues predicted that exclusion of grazing would inevitably increase biomass.[52] Biomass is fuel, it is not biodiversity.

Mitchell compared the low species richness on the clay floodplains of the Bogan against the high species richness of groundcover on Darling River sandplains:

> It has also its plains along the banks, some of them being very extensive ; but the soil of these is not only much firmer, but is also clothed with grass and furnished with a finer variety of trees and bushes, than those of the Darling. Yet in the grasses, there is not such wonderful variety as I found in those on the banks of that river. Of twenty-six different kinds gathered by me there, I found

only four on the Bogan, and not more than four other varieties, throughout the whole course. It appeared that where land was best and grass most abundant, the latter consisted of one or two kinds only, and, on the contrary, that where the surface was nearly bare, the greatest variety of grasses appeared.[53]

Nearly two centuries later, Price and Morgan compared studies of groundcover species richness on lighter and heavier soils. They concluded that "a common thread in these studies is that competitive dominance, whether it is by grasses, trees or shrubs, results in reductions in species richness".[54] If not for ignorance of our ecological history, academics might generally conclude that grazing can act an ecological analog for Aboriginal burning or natural flooding, by killing woody seedlings that no longer get burnt or drowned, and removing canopy of a few dominant grasses. This allows a multitude of more delicate herbs to establish[55] and maintains low fuel loads. Many areas of river red gum forest on the Murray have deteriorated into cherry scrub with dying trees[56] and bare ground since grazing was excluded. Biodiversity has declined and fuels have massively increased.

Away from the rivers, exceptionally herb-rich, dry woodlands of river red gum in Grampians National Park are suffering ongoing species decline since grazing was withdrawn because encroaching ti tree (*Leptospermum*) is choking out the herbs.[57] Most truly endangered species in Australia are threatened because remnants of their naturally open and usually grassy habitats, with scattered healthy, ancient trees and patches of bare ground, have been lost to thickening vegetation in the absence of grazing and/or burning. Examples are the mahogany glider and yellow-bellied glider in the wet tropics,[58] the Hastings River mouse[59] and eastern brown treecreeper[60] in northern New South Wales, the plains wanderer in south-eastern Australia[61] and the Tasmanian devil (see Chapter 6).

In contrast, where graziers have prevented woody thickening or cleared invading scrub, biodiversity has been maintained or enhanced. Calaby wrote of the area around Wallaby Creek in the Upper Clarence Valley: "The mammal fauna is the richest … in Australia. … in fact partial clearing in the rough grazing country has improved the habitat for several species of macropodids".[62]

Jansen's, Robertson's, Rowling's and Lunt's studies clearly showed that grazing kept fuels (shrubs, seedlings, saplings, large tussocks and herbs, and litter) low and discontinuous.[63] These results are typical of grazing studies and are typically reported as a negative impact – alleged loss of 'vegetation structure'. However a study by Dr. Dick Williams and colleagues – *Does alpine grazing reduce blazing?*[64] – sticks out from the ecological literature like the proverbial sore thumb. Williams and colleagues argued, against history and the ecological literature, that grazing cannot reduce fire severity.

Satellite images clearly show that the 2003 alpine fires fizzled out when they ran into the grazed area on the Bogong High Plains. Williams' study referred to these images but didn't use them. The images confirm eyewitness accounts by mountain cattlemen. Jack and Stuart Hicks had about 200 cows calving on the plains when the fires broke out, and they stayed with them for several days while the fires burnt around them.[65] However, Williams' colleagues estimated fire severity by measuring the diameter of twigs remaining on burnt shrubs in areas where the cattle mostly don't graze. They found that fire severity in the scrubs was unaffected by grazing[66] in nearby areas. Williams created a statistical model showing that grasslands only burnt where there were shrubs, but there was relatively little burnt in grazed shrubby grasslands and a relatively high proportion burnt in ungrazed shrubby grasslands. Obviously, grazing reduced blazing.

Incredibly, Williams and colleagues claimed that grazing "as a fire abatement practice is not justified on scientific grounds"[67] and co-author Carl-Henrik Wahren was one of the 'Concerned Scientists' that eventually persuaded the Victorian government to legislate against further trials of fuel reduction by cattle grazing in national parks. This was an important part of the 'science' underpinning the new legislation as evident in the following extracts from Hansard:[68]

The government's action on this important issue is a testament

to the efforts of dedicated environmental scientists ... There is
no question amongst the scientific community that cattle grazing
damages the environment and entirely fails to reduce fire risk. ...
There is no evidence to show that bushfire risk is substantively
reduced by allowing cattle to graze in the high country. Bushfire
spreads through the tree canopy. As far as I am aware, no cow
has ever existed that has been either genetically modified or
congenitally gifted enough to be able to climb a tree and graze upon
what is a major bushfire risk. ... There is overwhelming consensus
in the scientific community that cattle grazing is damaging to the
environment and does not achieve any significant reduction in
bushfire risk on a landscape scale. Clearly the cattle do not graze
on the flammable wood, bark and leaves that fuel bushfires in
the high country which can impact on people, property and the
environment.

Obviously grazing does not reduce trees, scrub, dead wood and litter,
but the scientific literature shows very clearly that it prevents these fuels
from accumulating, and it maintains an open and safe environment.
When fuels accumulate after grazing and burning have been excluded,
fire is necessary to reduce inedible fuels before grazing can work. It is a
bitter irony that green scientists and bureaucrats persuaded the Victorian
government to exclude grazing after the alpine megafires so that fuels
could 'recover', and attempted to prove that grazing doesn't work, before
lobbying to prevent scientific trials. Members in favour of scientifically-
based sustainable management expressed the following opinions:

I am disappointed that the Labor Party is playing the political game
... based not on scientific analysis but purely on political ideology.
Its decision will prevent any scientific studies from continuing. ...
Something that has been going on in Victoria since 1834 will be
outlawed just because of the ideological bent of a few members
of the government. ... To appease people in the inner city we
have a decision that will place Victorians in danger and remove
something that has been an important part of Victoria. ... This
bill is a product of the green nutbag left. It has nothing to do
with science. ... it is all about ideology ... We have seen that

when members of the left talk about issues they consider their own, issues in relation to which they like to quote from works of science, they go down the track of declaring the issue closed, saying, 'Science has decided this, so we will not be discussing it anymore'. My understanding is that science never closes. ... This bill is based on ideology and the Labor Party pandering to the Greens. The Labor Party is afraid of the Greens winning Labor's inner-urban seats, but that is going to happen anyway; the Greens are coming for Labor. In a desperate attempt to try to hold onto those inner-city seats the Labor Party is ... prepared to assign our heritage of cattle grazing to the scrap heap ... The tragedy is that as a result of the green religion and Labor's cynical, vote-grabbing nature, we have a situation where the Greens and Labor, in some sort of warped coalition, are prepared to lock up the bush and let it burn, and that is exactly what will happen.

Dick Williams reviewed *Losing Ground* – a book about the original political battles over grazing in the Alps half a century earlier.[69] He took from it a message quite different than I did:

> With the research of Alec Costin and colleagues, and the crucial 1957 Report by the Australian Academy of Science, there emerged substantial scientific evidence that grazing and catchment protection were incompatible. ... With its evidence that good scientific research, despite attempts to discredit it or dismiss it, was crucial in guiding government decision making, this book should be read by those who doubt the quality and relevance of scientific efforts – past, present and future – in high county [sic] land management in this country.[70]

In the book, author John Merritt referred in the following terms to a letter from the Secretary of Snowy Mountains Advisory Committee to Australian Academy of Science urging an "objective" study of grazing: "Raggatt was not thinking of 'impartiality' or 'disinterestedness' when he used the word 'objective' in his letter to the Academy's secretary; 'authoritative' and 'prestigious' would better describe what he was after. The report fitted both specifications". The blurb on the cover of Merritt's book suggests that:

The bitterness generated during the course of a 25 years dispute has survived until today, in some measure because of the partisan histories of the victors. John Merritt offers a more balanced account of the lessees' defeat. His focus is on arguments and political manoeuvrings, and he challenges posterity's charge that the lessees were selfish and unpatriotic.[71]

Every modern environmental battle in Australia since this first has been won by political manoeuvrings. But science continues to show that land managers wishing to maintain healthy and resilient landscapes (e.g., Fig. 7, p. 207) may choose grazing, burning and/or slashing. If they sit on their hands they will fail.

Endnotes

1 Norton 1887.

2 Close *et al.* 2008.

3 Gammage 2011.

4 Mitchell 1839, Vol. 1, pp. 238-9.

5 Curr 1883, p. 189.

6 Flinders 1814.

7 Mitchell 1839.

8 McIntosh *et al.* 2005.

9 Gammage 2011, p. 104.

10 Noble 1997.

11 Jurskis 2005.

12 Australian Academy of Science 2006.

13 Ibid.

14 Costin 1954.

15 Blair c1950.

16 Ibid.

17 Merritt 2007.

18 Gough 1994.

19 op. cit.

20 Ibid.

21 Costin 1954.

22 Merritt 2007.

23 Ibid.

24 Ibid.

25 Ibid.

26 Ibid.

27 Ibid.

28 Ibid.

29 Ibid.

30 Ibid.

31 Croke and Takken 2005.

32 Somerville *et al.* 2011.

33 Commonwealth of Australia 1995.

34 Concerned Scientists 2011.

35 Henderson and Keith 2002.

36 Tasker and Dickman 2004.

37 NSW Office of Environment & Heritage 2011.

38 Natural Resources Commission 2009, p. 9.

39 Jurskis 2009.

40 Jurskis 2005.

41 Jurskis 2005, 2009, 2011a.

42 Robertson and Rowling 2000.

43 Jurskis 2005, 2009, 2011a.

44 Lunt *et al.* 2007a.

45 Ibid.

46 Sturt 1838, Mitchell 1839, Curr 1883.

47 Wallis 1878; Manton 1895; Chesterfield 1986; Jurskis 2009, 2011a.

48 Price and Morgan 2008, 2010.

49 Jacobs 1955.

50 Stokes *et al.* 2010.

51 Lunt *et al.* 2007a.

52 Lunt *et al.* 2007b.

53 Mitchell 1839.

54 Price and Morgan 2010.

55 Price and Morgan 2008, 2010.

56 Sinclair 2006.

57 Price and Morgan 2008, 2009.

58 Stanton *et al.* 2014b.

59 Tasker and Dickman 2004.

60 Ford *et al.* 2009.

61 NSW Office of Environment & Heritage 2011.

62 Calaby 1966, p. 3.

63 Robertson and Rowling 2000, Jansen and Robertson 2001, Lunt *et al.* 2007a.

64 Williams *et al.* 2006.

65 Attiwill *et al.* 2009.

66 op. cit.

67 op. cit.

68 Parliament of Victoria 2015.

69 op. cit.

70 Williams nd.

71 op. cit.

10

MURRAY-DARLING MADNESS

Dead flies cause the ointment of the apothecary to send forth
a stinking savour
Book of Ecclesiastes 10:1

Nothing better illustrates the negative impacts of the wilderness mentality on our socioeconomic and environmental landscape than the current Murray-Darling Madness. Surveyor-General Mitchell recognised Australia's extreme climatic variability and the constraints it imposed on agricultural development. He foresaw the need for river regulation and irrigation if the colony was to develop into a new nation:

Agricultural resources must ever be scanty and uncertain, in a country where there is so little moisture to nourish vegetation. We have seen, from the state of the Darling, where I last saw it, that all the surface water flowing from the vast territory west of the dividing range, and extending north and south between the Murray and the tropic, is insufficient to support the current of one small river. The country southward of the Murray is not so deficient in this respect, for there the mountains are higher, the rocks more varied, and the soil, consequently, better ; while the vast extent of open, grassy downs, seems just what was necessary, for the prosperity of the present colonists, and the encouragement of a greater emigration from Europe.

Every variety of feature may be seen in these southern parts, from the lofty alpine region on the east, to the low grassy plains, in which it terminates on the west. The Murray, perhaps the largest river in all Australia, arises amongst those mountains, and receives in its course, various other rivers of considerable magnitude. These flow over extensive plains, in directions nearly parallel to the main stream, and thus irrigate and fertilise a large extent of rich country. Falling from mountains of a great height, the current of these rivers is perpetual, whereas in other parts of Australia, the rivers are too often dried up, and seldom, indeed, deserve any other name than chains of ponds ...

The grassy plains which extend northward from these thinly wooded hills, to the banks of the Murray, are chequered by the channels of many streams falling from them, and by the more permanent and extensive waters of deep lagoons. These are numerous on the face of the plains near the river, as if intended by a bounteous Providence, to correct the deficiencies of too dry a climate. An industrious and increasing people, may always secure an abundant supply, by adopting artificial means to preserve it, and in acting thus, they would only extend the natural plan, according to their wants...

The Murray, fed by the lofty mountains on the east, carries to the sea a body of fresh water, sufficient to irrigate the whole country [behind Cape Northumberland – the Mallee and Wimmera Regions of Victoria], which is, in general, so level, even to a great distance from its banks, that the abundant waters of the river, might probably be turned into canals, for the purpose, ... of supplying deficiencies of natural irrigation.[1]

The Murray-Darling's first small irrigation scheme commenced in the 1870s near Kerang on "the grassy plains ... chequered by channels" that Mitchell had described. A weir on the Goulburn River was constructed during the 1880s. The Federation Drought was the impetus for the construction of Burrinjuck Dam on the Murrumbidgee River. Construction commenced in 1907, but various problems including the First World War delayed progress so that the dam wasn't completed until 1928. It was upgraded in 1957 and 1994. On the Murray, Torrumbarry Weir was finished in 1929, the original Hume Reservoir in 1936, and Yarrawonga Weir in 1939. Stevens Weir was completed on the Edward in 1935 and Mulwala Canal connected Yarrawonga Weir to the Edward River in 1941. The capacity of Hume Reservoir was doubled by 1961.

The Snowy Mountains Hydroelectric Scheme was constructed between 1949 and 1974 by more than 100,000 people from over thirty countries. Sixteen dams, seven major power stations and hundreds of kilometres

of tunnels and aqueducts were built to supply clean renewable energy
to all capital cities from Brisbane to Adelaide. Additional water was
provided for irrigation in the Murray Basin by diverting water which
was previously lost to the ocean via the Snowy River.[2]

At this stage I must disclose a personal interest. My father Vytautas,
shouting most authoritatively in Russian, bluffed his way through an
advance guard of Asian conscripts during Stalin's invasion of Lithuania
in 1944. He escaped firstly to Austria, and briefly worked with the U.S.
occupying forces in Bavaria before being recruited to assist Australia's
post-war development. I lived my first few years at Adaminaby Dam, the
construction camp for what is now called Lake Eucumbene. I am proud
of this connection with the greatest sustainable development Australia
will ever see.

The Snowy was arguably the birthplace of multiculturalism in
Australia. Alpine graziers (immortalised in *The Man from Snowy River*[XVI])
discovered gold at Kiandra in 1859. At the height of the 'rush' there were
15 hotels and 30 stores. Norwegian miners introduced skiing in 1861, and
a skiing club held competitions in the 1870's. Three Mile Dam was built
by Chinese miners in 1882 and connected to the goldfields by miles of
water race. Sluicing and dredging continued at Kiandra until 1905. Some
of the more recent interactions amongst the people of many different
nations working on the Snowy Scheme are now also legendary and are
the subject of an interesting book by Noel Gough who started with the
Snowy Mountains Authority at Three Mile Dam in 1950.[3]

The Queen's Lookout, an enclosed viewing platform with associated
facilities, was constructed above Lake Eucumbene for an official opening
of the Scheme by Her Majesty Queen Elizabeth II in 1963. This item of
cultural heritage along with most of Kiandra was later demolished, and
the historic Jounama arboretum was clearfelled by the National Parks
and Wildlife Service as they increasingly restricted public access to much
of Kosciuszko National Park, reduced prescribed burning, and allowed
pests and weeds to flourish in the absence of grazing and burning.

XVI A famous poem by A.B. (Banjo) Paterson, and later a movie.

However the Service is not entirely opposed to human economy. Whilst they were creating wilderness in most of the park, they were planning the Skitube development. This is a nine kilometre rack railway, including six kilometres of tunnels and three stations, completed in 1988.

At the same time, many city-based 'snow bunnies' of the wilderness persuasion were saving the Franklin River and electing the Hawke Government. The consequent sweeping powers of the Federal Government to intervene on environmental matters, together with climate hysteria, set the scene for the current Murray-Darling Madness. Queensland, New South Wales, the Australian Capital Territory, Victoria and South Australia share the resources of the Murray-Darling Basin. Naturally each Government thinks that the other States are getting a better deal. Add conflicting demands for irrigation, power generation, flood mitigation and environmental flows, and it is obvious that there will be ongoing tension.

The tension would be considerably reduced by dispensing with the strange concept of environmental flows. For example, the Millenium Drought prompted governments to develop an eighty million dollar scheme to divert environmental flows into the unnaturally dense Koondrook-Perricoota forests to water trees. Major floods in 2010, 2011 and 2012 watered the forests before construction was finished and the first 'environmental flows' were delivered in 2014.[4] As Mitchell observed, there is extreme long-term variability in flows with climate, and long periods of wet or dry. Irrigation schemes "only extend the natural plan".[5]

But there is a fly in the ointment. During the 1930s the South Australian Government built a number of sea dykes to block the channels that converge on the mouth of the Murray. This effectively converted Lake Alexandrina from an intermittently closed tidal estuary into an artificial freshwater storage dependent on flows from the Murray-Darling Basin. A thriving mulloway fishery was destroyed in the process. This feat of engineering certainly did not "extend the natural plan". The storage is used for irrigation and domestic water supply. More recently, the problem has been compounded by the development of housing estates, resorts and marinas on naturally flood prone lands around the artificial storage. Constant dredging is required to keep the river mouth open.[6]

Naturally, South Australia doesn't want to be seen to be decimating agricultural production higher in the Basin to support dwindling agriculture and increasing urbanisation around their huge artificial storage. So the Government pretends that this was a freshwater lake system. They have avowed a commitment to maintaining the Lower Lakes as a 'natural' system. Green academics from other states have joined in a conspiracy to misrepresent our ecological history. Apparently they prefer to see unnaturally stable and high populations of their favorite wildlife rather than reliable production of food for people. Hence the Federal Government's ten billion dollar plan[7] to reduce irrigation, buy back licences, and send more water down to the mouth and out to sea.

Professor Richard Kingsford of the Australian Wetlands and Rivers Centre has been counting birds from aeroplanes since 1986 because:

> they give me a wonderful opportunity to move and follow them across large parts of Australia … to do that over more than a quarter of a century has meant a real perspective … about how our continent responds … And its pretty arrhythmic because we have these boom and bust periods. Floods are absolutely critical.[8]

After bird surveys in 2011 The (Brisbane) *Courier Mail* reported that:

> Survey leader and director of the Australian Wetlands and Rivers Centre at the University of NSW, Richard Kingsford, said yesterday he was astounded by the numbers. … Prof Kingsford said the richness of breeding species was also high and the floods had seen a turn around in a 30-year decline.[9]

But during waterbird surveys in 2014, Kingsford told Bill Birtles on *PM* that there was little water in any of the major lakes in western Queensland. He said "The overall pattern that we've seen over those 30 years is for a decline in waterbird numbers".[10] This was before he'd counted the birds in the south where 'environmental releases' have occurred, and only three years after he counted astounding numbers of birds.

Kingsford claims that populations of waterbirds don't bounce back as much as they used to because of river regulation.[11] This is unbelievable because river regulation cannot stop big floods, but it does stop many wetlands from drying out completely. One wonders whence came all the waterbirds when Lake Eyre completely filled in 2010 after a decade of drought in the Murray. Kingsford's motivation appears to be a mixture of naturism and xenophobia:

> I think this is a big challenge for humanity. We've actually got to work out how we can have less of a footprint, less of a water footprint. In Australia, a lot of what we get out of our water is exported, and so making sure the rest of the world raises their standard of living on the backs of, or using the water from our rivers, I think is worrying. Ultimately it comes down to what sort of impact are we having on the environment, and what sort of environment do we want to leave for you know, our future generations?[12]

Lake Alexandrina was a tidal salt water estuary when Sturt arrived in 1830, and he saw a seal at its southern end.[13] Perhaps it is this more natural environment that we should leave for future generations together with the food and prosperity obtained by "extending the natural plan" as foreshadowed by Mitchell.

Unlike the Murray, the Snowy River retains its intermittently closed tidal estuary, its diverse native fishery and its value as a Site of Geomorphological Significance (according to the Geological Society of Australia). Reduced flows after diversions to the Murray are still sufficient to keep the estuary open to the ocean most of the time, even though the river mouth naturally migrates considerable distances back and forth along the beach. The diversions of water to generate clean energy and support food growers in the Riverina have not changed the character of the estuary.

But environmentalists begrudge these diversions, and have successfully lobbied for environmental flows.

New South Wales' Office of Water explained how environmental flows in the Snowy will improve the environment:

the primary ecological objective would be achieved by storing and then releasing sufficient volumes to provide flushing flows. These high flows facilitate the improvement in the physical condition of the in-stream habitat by scouring and transporting sediment (i.e., primarily, sand, silt and clay). This primary ecological objective was identified as a high priority for the environment during the Snowy Water Inquiry. These high flows are important to improve river health. It is expected that as the river physically responds over time to the higher flows, the focus of the recovery process will transition to meet other secondary ecological objectives.[14]

The obvious catch is that these 'flushing flows' move relatively small amounts of sediment further down towards the East Gippsland Plain where it settles out when the gradient of the channel levels out. This may marginally improve the appearance of the river around Dalgety, but major floods are required to move any substantial amounts of sediment and flush it out to sea. Jindabyne Dam, like Hume Dam on the Murray and Burrinjuck Dam on the Murrumbidgee, doesn't prevent big floods. For example, the Snowy River flooded to a height of 7.8 metres at Orbost in 2012.

Environmental flows will not address the major environmental problems on the upper Snowy, including soil erosion; feral willows, blackberries, foxes, rabbits and trout; and chronic decline of eucalypts. These problems did not suddenly arise when Jindabyne Dam was commissioned in 1967 as part of the Snowy Mountains Scheme. Providing for environmental flows in the Snowy River cost taxpayers about half a billion dollars between 2002 and 2014. It also cost 150 gigawatt hours per annum of clean renewable energy, which is sufficient to power about 18, 000 homes, and equates to over 100,000 tons of CO_2 emissions from a thermal power station.[15]

Every living species in Australia has persisted through climatic extremes. Fluctuating populations are natural. Habitats expand and contract. Environmental flows can do nothing but dampen fluctuations. They cannot stabilise populations of ephemeral species such as waterbirds or eliminate recurrent climatic stress on long lived sedentary species such as river red gums. River regulation has destroyed some wetlands and created some new ones. It has drowned some woodlands and created some new forests. We should enjoy the new food, timber and clean energy resources that we have created. The waterbirds can look after themselves, but the new forests can't, even if they're watered.

Endnotes

1 Mitchell 1839.
2 Snowyhydro nd.
3 Gough 1994.
4 State Water nd.
5 op. cit.
6 Marohasy 2012.
7 Murray-Darling Basin Authority nd.
8 Catalyst 2012.
9 Williams 2011.
10. Colvin 2014.
11 Catalyst 2012.
12 Ibid.
13 Sturt 1833.
14 NSW Office of Water 2010.
15 NSW Office of Water 2010, Boco Rock Wind Farm nd.

11

SAVING FORESTS THAT NEVER WERE

The history of trees and forests is writ large in their architecture

Attributed to Maxwell Ralph Jacobs by F.D. Podger

The Millenium Drought and controversy about the Murray provided an opportunity for the disintegrating Labor Government of New South Wales to deliver its final gift to the Greens in 2010. The bribe didn't pay off and the Government was ousted in 2011, partly because people were beginning to notice the socioeconomic havoc wrought by governments gutting rural communities to buy Green votes in the cities. Unfortunately, by this time some communities such as Mathoura in the Riverina had been virtually destroyed.

Fig. 8 Curr saw this tree in 1843: "there were no blacks there at the moment ; but some camp fires smouldering under the shade of a spreading tree, and the bags and nets which hung from its branches, showed that they were not far away".
(Courtesy of NSW Department of Primary Industries)

The journals of Oxley, Sturt and Mitchell, Curr's *Recollections*, historical maps and photographs, reports by nineteenth century foresters, the structure of remnant pre-European stands, and the form of ancient trees leave no doubt that river red gum 'forests' were lines and narrow strips of trees along rivers, anabranches, flood runners[XVII] and creeks, around lakes or billabongs, and in river bends. Frequently flooded plains carried reedbeds, and higher, less frequently flooded plains carried grasslands or open woodlands of red gum (e.g., Fig. 8, p.255) or box (*Eucalyptus*).[1] Large numbers of Aborigines used these areas and burnt the reedbeds and grassy woodlands. They lit fires for hunting, for corroborees, and for wars with other tribes. Wood fuelled their ovens and campfires, and supported the sheets of bark that clad their huts.[2]

Red gum seedlings had little chance of surviving fires or trampling. Some small scrubs had developed after smallpox decimated the tribes in 1789.[3] But extensive scrubs didn't grow until after squatters such as Curr had disrupted Aboriginal culture from the mid-nineteenth century. Wet conditions and high floods during the 1870s established a dense growth of red gum in many former grasslands and woodlands, turning them into scrubs. Foresters, lessees and jobless refugees from cities thinned the scrubs during the 1890s depression to grow grass and timber.[4] After river regulation changed flooding patterns from the early-twentieth century, scrubs invaded additional areas of reedbeds and swamps.[5] A century after red gum forests came into being, wilderness fanatics from the cities began campaigning to save these 'iconic natural forests' from the communities that had created and maintained them.

As Walter Starck succinctly put it:

> environmentalism has redefined the fundamental concept of being a stakeholder. Despite having nothing invested and with no risk to themselves, environmental non-government organisations [NGOs] have managed to claim the status of stakeholders in remote matters and be accorded an equal voice to those whose entire lives, livelihood and assets are being affected.[6]

XVII Ephemeral effluent creeks or minor anabranches.

Actually NGOs have been accorded a bigger voice than real stakeholders. In addition to the locking up of man-made forests to 'protect' natural heritage, this is evident in the Nature Conservation Council's privileged position on New South Wales' Bushfire Coordinating Committee.

Melbourne Greens achieved the locking-up of Victorian red gum forests in 2008 and Sydney Greens were encouraged. The dysfunctional New South Wales Government gave them an opportunity to impose their will on regional communities. Andrew Fraser MP told Parliament that Environment Minister Frank Sartor said at Gulpa sawmill in Deniliquin: "Let me give you a lesson in politics. The Greens hold 15% of the vote. And if we are to stay in power, we must hold their preferences to maintain city seats. They want a significant national park in red gum".[7] I alluded to this in my evidence to the 2012 New South Wales Inquiry into Land Management. The Honourable Luke Foley, shadow environment minister, asked whether I was aware that Sartor had appeared before the Inquiry and "directly addressed those claims", and whether I was aware that "he has repeatedly denied those claims". I answered truthfully that I was not aware.[8]

Later I read the transcript of Sartor's evidence to the committee and found that he had merely stated that "Some of the reports about who said what to whom, are totally wide of the mark".[9] In any case, my involvement in the Natural Resources Commission (NRC) assessment of the river red gum had left me in no doubt that it was flawed, and I also had the misfortune to witness conservation bureaucrats and Aborigines celebrating their 'win' at dinner in a Deniliquin hotel, in August 2009, months before any reports had been released or announcements made. At that time my colleague, Dr. Huiquan Bi, and I were engaged in fieldwork that was intended to inform the assessment. In the event it was an expensive waste of time.

Sartor acknowledged that local communities would suffer significant socioeconomic dislocation[10] as a result of Government delivering intangible 'benefits' to city-based wilderness enthusiasts. Environmental bureaucracies paved the way for these political decisions by misrepresenting the ecological history of river red gum forests, and Professor of Forestry at ANU, Peter Kanowski, chaired NRC's Technical Review Panel. However, after the last nail was driven into the coffin of Tasmania's native forest industry, he disparaged the "offstage deals and go for broke strategies" of the "dominant actors" in Tasmania's "tragedy". Kanowski suggested that NGOs and big business had conspired in this theatre.[11] I don't think it matters whether NGOs, big business, bureaucracy or academia 'wheel out the bodies', it's still a tragedy.

The *Final Assessment Report* of the NRC claimed that:

> The region's floodplains, and the river red gum forests they have
> supported for thousands of years, are an integral part of the natural
> landscape. ... But the river red gum forests are in decline across
> the region. As we've built dams and weirs over the last 120 years
> and diverted river flows to irrigation, the natural flows and floods
> of the river have been progressively restricted. Without flooding,
> the red gum forests can't regenerate and support the ecology
> and forestry industries which depend on them. The impacts on
> communities, land use and the natural environment are predicted
> to get more severe under climate change.[12]

The Commission stated that "ringbarking was carried out to open up pastures ... a thick understorey of natural regeneration developed in newly cleared areas" after the 1870s floods.[13] In fact, Curr who first grazed his stock in the reedbeds, grassy plains and woodlands had exalted that "this country possessed from the first, over a great portion of its area, the inestimable advantage of being ready for immediate use without the outlay of a sixpence".[14] Disruption of Aboriginal burning allowed extensive scrubs to develop. Foresters thinned scrubs and ringbarked old woodland trees to create the new forests.[15]

However, NRC ignored the evidence of Oxley, Sturt, Mitchell, Curr and other early settlers as well as the foresters who did the work. They claimed that "it is difficult to draw conclusions about the linkages between the use of fire and forest structure at the time of European settlement".[16] One of the references listed to support this gross distortion of history contains a survey map from 1848 and a detailed, quantitative description by a forester in 1895 which prove beyond doubt[17] that the forests developed after Aboriginal burning was disrupted.

Had the Commission presented a true ecological history, lack of 'regeneration' with reduced flooding may have been seen in a positive light, and they might have recommended logging and burning to restore more natural woodlands. Even taking the report at face value, it makes no sense to reserve the well watered forests above the Barmah Choke[XVIII] and to continue producing timber from drier downstream forests with 'regeneration difficulties' and to spend 80 million dollars digging a ditch to water these artificial forests.

Having paid compensation for destroying local industries that were thinning dense forests, Victoria and New South Wales agreed to jointly trial 'ecological thinning' to waste on small areas. This will provide negligible employment, waste renewable energy, mess-up semi-natural habitats, interfere with human access and recreational values, create fire hazards, and burden taxpayers. In 2012 NSW Office of Environment and Heritage and Parks Victoria published a plan for the thinning trial, which stated that:

> The duration and frequency of floods would have acted to limit widespread recruitment of high density stands. In concert with altered flood regimes, management for commercial timber extraction has altered these natural processes, resulting in widespread establishment of high stem density stands and a paucity of large trees in the landscape.

So the green bureaucrats have flooding or lack of flooding preventing recruitment of red gum according to their want. In fact, most of the red gum forests established soon after Aboriginal burning was disrupted. Forester Manton reported on New South Wales' Central Murray forests in 1895 as follows:

> there are two matured red-gum trees to the acre fit for saw-mill

---

XVIII A constriction of the river channel that causes flood waters to spread widely above the choke.

purposes; trees of full growth, but valueless by reason of their
being hollow, spongy, and winding growth, at about eight to the
acre; young, vigorous, and healthy trees, varying from16 in. to 20
in. diameter, may be reckoned at seven to the acre; while on large
areas there is a dense growth of young trees numbering in places
over 2,000 to the acre.[18]

In other words, there were only twenty five trees per hectare before
European settlement, of which five were good for timber. Seventeen
young trees per hectare established soon after European settlement in
1842. They grew to 40 or 50 cm. in diameter over as many years. Up to
five thousand saplings per hectare established after major floods in the
1870s. This report leaves absolutely no doubt about how and when the
river red gum woodlands turned into forests.

Some additional areas of forest established in former reedbeds and
swamps consequent to reduced flooding in the twentieth century. One of
these reedbeds, described by Curr in 1843, was invaded by red gum after
Hume Reservoir reduced floods. Sixty years later a university student
portrayed the new forest as 'old growth'.[19] He measured windrows of
debris that had been cleared off a management trail to set the benchmark
for 'natural' loads of fallen timber.[20]

Contrary to the assertions of conservation bureaucrats, "management
for commercial timber extraction" has consistently reduced stand densities
in the new forests, producing social, economic and environmental
benefits. Ecological thinning to waste is a bad joke, and the rationale
for wasting the resource is the false benchmark for fallen timber. Heavy
commercial thinning and frequent mild burning would be the sound
ecological approach to restore and conserve diverse woodlands.

In any case, during the run up to the Victorian elections in November
2014, the Napthine Government decided to withdraw from the thinning
trial. The decision was announced by Friends of the Earth on 23rd
September in the following terms:

A letter from the Victorian Department of Environment and
Primary Industries, received by Friends of the Earth today,

has confirmed that Victoria has decided to withdraw from the proposed trial and will not be seeking approval to proceed with the trial under the Federal Environment Protection and Biodiversity Conservation [EPBC] Act.

This decision on top of the 'saving of the red gum' in two States, the micromanagement provisions of the Commonwealth EPBC Act, and all the other bad 'environmental' decisions by State or Federal governments going back to the Franklin Dam debacle illustrates that socioeconomic and environmental chaos results when major parties go 'shopping' for green support before elections. The way it was announced confirms the insidious political influence of greens. A fundamental flaw has appeared in Australia's political system. Queensland was great under Joh Bjelke-Petersen's infamous 'gerrymander' because genuine regional stakeholders had a say, and their influence was transparent. Now genuine stakeholders are disenfranchised throughout Australia by the disproportionate influence of unelected greens in city electorates.

A member of the Technical Review Panel for the NRC 'assessment' of river red gum in New South Wales took the opportunity of his collaboration in the flawed process to write a book about the "Ecology and History of the River Red Gum".[21] I have selected a few quotations to illustrate the flaws in both the process and the book. Dr. Matthew Colloff quoted Curr to headline his chapter about fire in river red gum: "I refer to the *fire-stick*; for the black fellow was constantly setting fire to the grass and the trees." Colloff claimed that "This statement by Edward Curr regarding the Aboriginal use of fire in river red gum forest has been widely misquoted and misused ... Curr was not talking specifically about these forests" even though Curr referred to "almost every part of New Holland".

Colloff claimed that Curr "speculated" about the effects of burning, and he used a quote from Ron Hateley writing 130 years after the event to 'demonstrate' that Curr's **observations** were "obviously wrong". Colloff made contradictory statements that fire's "role in the management of river red gum forests remains unclear" and that "the use of fire by Aboriginal people to manage the structure and productivity of particular vegetation communities for their needs, embodied in the concept of fire-stick farming advanced by Rhys Jones, is generally accepted and widely acknowledged".

Curr advanced the concept of firestick farming a century before Rhys Jones, as evident from the headline to Colloff's chapter. He developed the concept whilst occupying the riverine floodplains. Colloff wrote of Gammage's book:[22] "among the great many quotes I found none about the deliberate burning of riverine floodplain forests". The very obvious reason for this is that the "riverine floodplain forests" did not exist when Aborigines were managing the floodplains. Anyone reading Curr's book with an open mind would have no trouble grasping this obvious fact.

Gammage and Colloff both used this quote from Curr: "A sea of reeds, of several miles in extent … by patches and strips of different hues and growth, in accordance with their ages and the periods at which they had last been burnt". This shows that the Aborigines did indeed burn the floodplains. These reedbeds are now the Barmah Forest[23] which Colloff claims has occupied the floodplain for 3000 years. This is where the Victorian thinning trials were to occur before they were cancelled to appease greens.

Under Aboriginal management there were narrow strips of red gum along riverbanks and scattered trees in reedbeds and grassy plains. Indeed, Colloff provided a quote from Sturt showing unequivocally that Aborigines burnt stands of river red gum:

> We found the country on both sides of the river free from reeds but heavily timbered … the reeds had been burnt by the natives and in burning [they] had set fire to the largest trees and brought them to the ground. [Colloff's editing.]

Colloff also used another quote from Curr about these floodplains: "On closer examination we also noticed several patches of good country, such as we had just left, and that the whole could be made available, with little trouble by burning". He stated that:

> This observation begs the question of how burning would make

the country more 'available' if the country was already 'open'. I
suspect Curr was referring to the use of fire not to improve access
for sheep but to enhance the palatability of grasses.

Curr and his brother had made their "closer examination" by
climbing "a gnarled and spreading gum-tree, from the branches of
which a view of the neighbouring country might be obtained". This old
spreading tree was in a "sea of reeds" through which they "forced our
horses, with considerable difficulty".[24] An open-minded reader would
have no trouble understanding that Curr intended to burn the reedbeds
to provide access and feed for his stock. Colloff also quoted Sturt on
the difficulties of access through old reedbeds.

Curiously, he claimed that "Sturt mentions the Barmah-Millewa was
'heavily timbered'."[25] Here is an actual quote from Sturt as he entered the
area that is now forest:

> we gradually advanced upon a region of reeds and swamps but
> even in the midst of these, the grassy flats, though more heavily
> and more closely timbered than similar spaces on the rivers to the
> northward, were luxuriant and extensive.[26]

It is clear that reeds and swamps were the dominant feature, and that
the higher plains carried grassy open woodlands, albeit with more trees
than those on the Murrumbidgee and the Lachlan. There were about
twenty trees per hectare in red gum woodlands at Barmah-Millewa.[27]
Sturt described what is now Barmah Forest as a "vast marsh".[28] Colloff
and NRC had access to all this historical information and chose to either
ignore it or misinterpret it. Colloff's book includes an historic map
(attributed to Surveyor Townsend in 1852, but apparently drawn by a
Mr. Boyd in 1848) showing that what later became the Millewa, Moira
and Gulpa Island red gum forests were mostly reedbeds and swamps.[29]
Colloff suggest that a label "Gum Swamp" on this map denoted river
red gum forest. Ironically river red gum takes it scientific name from
Camalduli in Italy where Mussolini later planted it to dry out the Pontine
Marshes.

According to Colloff "the view that forests were virtually absent before
European settlement ... is puzzling" because of the large quantities of

timber produced soon after. Forester Manton's report[30] indicates that the timber was produced from about five big old trees per hectare. This is why there are no forests of stumps as would be expected if forests had been felled. Jacobs reputedly said: "the history of trees and forests is writ large in their architecture". In the new forests, old relict trees on the high banks of rivers and along runners, in river bends and around billabongs are tall and clean-boled. But old trees and stags on floodplains are short with widespreading branches (e.g., Fig. 8, p. 225). There are many old stumps along rivers and runners, there are few on floodplains.

Lines of forest grew where red gum had easy access to shallow groundwater so that seedlings could quickly grow into saplings that were immune to destruction by low-intensity fires. Colloff referred to nineteenth century reports by Victorian foresters to emphasise the early timber yields. These reports also show that Gunbower forest was "originally rather thinly timbered", that the area of river red gum at Gunbower increased by 7,000 ha after European settlement, and that the potential sawlog yield of the new Barmah Forest is an order of magnitude higher than what was cut from the old red gum stands.[31]

I summarised measurements from 15 pre-European stands of river red gum on the flood plains of the Murray and Murrumbidgee Rivers. One of these featured in the NRC report as an example of a healthy ecosystem (Fig. 7, p. 207). There were between 6 and 20 trees per hectare in these stands. The average was 11, and all the trees were very large. They are nothing like the new forests with ten times as many trees, even after commercial thinning.[32] One of Colloff's colleagues on the Technical Review Panel has published data from other eucalypt woodlands on the Western Plains showing pre-European densities of 20 trees per hectare. All this information about vegetation change was known to Colloff and the Natural Resources Commission. They cited my article summarising the historical evidence on red gum stand structures,[33] and Colloff also cited my later publication dealing with woodlands more generally.[34] But they didn't discuss any of this evidence. It was dismissed out of hand because it allegedly came "from a forestry perspective" and "was limited to New South Wales and three expeditions".[35] It actually came from explorers, squatters, foresters, newspapers and measurements in New South Wales and Victoria.

Colloff claimed that:

> The view of ecologists, who regularly have to address recent environmental change as part of their research, tends to be that historical evidence is problematic because it is open to interpretation and that debate over differences may be unprofitable and contentious. Ecologists would rather collect historical data empirically where possible.

But he misinterpreted unequivocal historical information from Sturt, Mitchell, Curr, Manton and others about the nature of vegetation in precisely known locations and uncritically dismissed empirical data, making it clear whence arises the contention. According to Colloff, the NRC was "repeatedly accused of ignoring local views and of incompetence and corruption".[36] It is clear that the NRC ignored soundly based local views formed on unequivocal historical evidence. The question of competence is very much open.

Premier Nathan Rees announced the decision to 'save' the Millewa forests on 3rd December 2009. The Wilderness Society welcomed the "fantastic outcome for the environment". NSW National Parks Association stated that "We look forward to a similar announcement across the remaining River Red Gum Forests when the Natural Resources Commission releases their final recommendations for the forests' future on December 21". Premier Rees was deposed by Kristina Keneally the same day. Keneally announced the rest of the plan for socioeconomic and environmental destruction in March 2010. Greens held the balance of power in the senate and didn't agree with the plan to phase out logging over five years. The final dirty deal was done in May, after the government substituted blood-money for phaseout of logging.

It was reported that:

> Bob Brown congratulated the NSW government … Premier Keneally sidestepped a question that the deal was done to secure preferences from the Greens in the upcoming March 2011 state

election. 'This is about doing the right thing for what is a nationally and internationally significant forest,' she told reporters in Sydney. 'It's about protecting threatened species, it's about protecting ecosystems.'[37]

Major floods in 2010, 2011 and 2012 exposed the weakness of alarmist arguments used by NRC to deliver the preconceived outcome. But I don't really understand why the Government saw a need to disguise the political decision as a scientific one, nor why academics assisted. No-one was fooled because they'd seen it all before.

In 2005, just before he jumped the sinking ship that was the NSW Labor Government, Bob Carr broke an earlier commitment to rural communities and gifted the Greens most of the Pilliga Scrub. Unfortunately, the ship sank very slowly after opposition leader John Brogden publicly referred to Carr's Malaysian-born wife as a mail-order bride.[38] Together with some other unfortunate events caused by stress, this led to his resignation as the leader of the opposition party.[39] An independent won the consequent by-election, whilst support for the Labor Government ceased to falter under new Premier Morris Iemma. In 2007 Iemma won the 'unwinnable' election, and the opposition were unable to fulfil their election promise to reverse Carr's lock-up of 350,000 hectares of productive man-made forest.

Cypress is extremely drought-tolerant and grows on gently undulating country lacking permanent watercourses. Thus Carr didn't need to wait until Al Gore's convenient untruth inflamed Murray-Darling Madness before 'saving' cypress forests from the communities that created them. Post-European cypress scrubs had been thinned to provide timber, grazing and biodiversity in southwestern New South Wales since the 1890s depression and in the Pilliga since the 1930s depression.[40] Carr's decision created another depression in rural communities around the Pilliga. More megafires such as the 2013 Coonabarabran disaster[41] are inevitable with the dilution of human economy after active multiple use management was replaced by wilderness style neglect.

Here also, green academics collaborated with green bureaucrats to rewrite ecological history. In 2011 they published a paper asking

"How do slow-growing, fire-sensitive conifers survive in flammable eucalypt woodlands?".[42] The paper portrayed dense post-European cypress scrubs as natural ecosystems and concluded that suppression of ground fuels is a natural survival mechanism. It warned that this mechanism wouldn't work in future when human caused global warming inflames frequent firestorms. According to Flannery's taxpayer funded propaganda, the 2013 megafires vindicated them.[43] But graziers originally took up their holdings in the Pilliga because it was grassy woodland. The correct answer to the rhetorical question posed by the academics is that cypress is not sensitive to mild fires in its natural habit as an open grown tree in grassy woodland where fire storms cannot develop (see Front and Back Cover).

As outlined by James Noble in *The Delicate and Noxious Scrub*,[44] CSIRO and State governments have spent countless millions of dollars researching potential solutions to the scrub problem in rangelands. River red gum and cypress forestry turned environmental problems into socially, economically and environmentally sustainable industries. The deliberate dismantling of these industries, with the communities that depend on them, and the reinstatement of the problems says much about green ideology and the wilderness mentality. None of it is good.

Endnotes

1 Oxley 1820, Sturt 1833, Mitchell 1839, Wallis 1878, Curr 1883, Manton 1895.

2 Oxley 1820, Sturt 1833, Mitchell 1839, Curr 1883.

3 Curr 1883, Flood 2006.

4 Manton 1895, Donovan 1997.

5 Chesterfield 1986; Jurskis 2009, 2011.

6 Starck 2014.

7 Parliament of New South Wales 2010.

8 General Purpose Standing Committee No. 5 2012a.

9 General Purpose Standing Committee No. 5 2012b.

10 Ibid.

11 Kanowski 2012.

12 Natural Resources Commission 2009.

13 Ibid., p. 24.

14 op. cit.

15 Manton 1895.

16 op. cit. p. 24.

17 Donovan 1997; p. 18, pp. 96-8.

18 Manton 1895.

19 Robinson 1997.

20 Jurskis 2011a.

21 Colloff 2014.

22 Gammage 2011.

23 Curr 1883, pp. 169-71; Chesterfield 1986, Fig. 3.

24 Curr 1883, pp. 171-2.

25 Colloff 2014, p. 248.

26 Sturt 1838.

27 Manton 1895.

28 op. cit.

29 Donovan 1997, Fig. 6; Colloff 2014, Fig. 5.4.

30 op. cit.

31 Wallis 1878, Tucker *et al.* 1899, Jurskis 2009.

32 Forestry Commission of N.S.W. 1985, Appendix 7a; Jurskis 2009.

33 Jurskis 2009.

34 Jurskis 2011a.

35 Colloff 2014, p. 104.

36 op. cit.

37 Middleton 2010.

38 Pearlman and Clennell 2005.

39 Brogden nd.

40 Rolls 1981, Jurskis 2009.

41 General Purpose Standing Committee No. 5 2015.

42 Cohn et al. 2011.

43 Happs 2013.

44 Noble 1997.

12

RAINFOREST, WET SCLEROPHYLL AND SCRUB

Oils ain't oils Sol

Gangster reprimanding his getaway driver in a motor oil advertisement

When I started as a professional forester in 1978, rainforests were 'flavour of the month'. Green media such as the Australian Broadcasting Commission were spruiking dire predictions that the Amazon jungle would all be gone within twenty years and they were constantly repeating claims about the fragility of rainforests. On the day I arrived in Casino to report to the Forester in Charge of Casino West Subdistrict, Assistant Clerk Darrel Couch told me "we've got no jobs going here mate". I soon found out that local professional staff were in Kyogle attending hearings of the State Pollution Control Commission, which culminated in the conversion of the rainforests of Wiangaree State Forest into the Border Ranges National Park.

Encouraged by winning the Border Ranges, greens mounted a campaign to save Terania Creek "rainforest" from logging. Their campaign initially focused on the danger of log trucks sharing the narrow access road with a school bus. When loggers offered to work around the bus timetable, greens shifted their focus to the 'fragile' rainforest. Foresters agreed to exclude rainforest from logging. Greens' focus shifted to the rainforest subcanopy in wet sclerophyll forest. This provided impetus for some research into the relationship between rainforest, sclerophyll forest, environment and fire by John Turner who was then a Senior Research Scientist with NSW Forestry Commission.

Turner's work showed that rainforests grow in spaces defined by the interaction of climate, topography, soils and firestick. As soil depth and

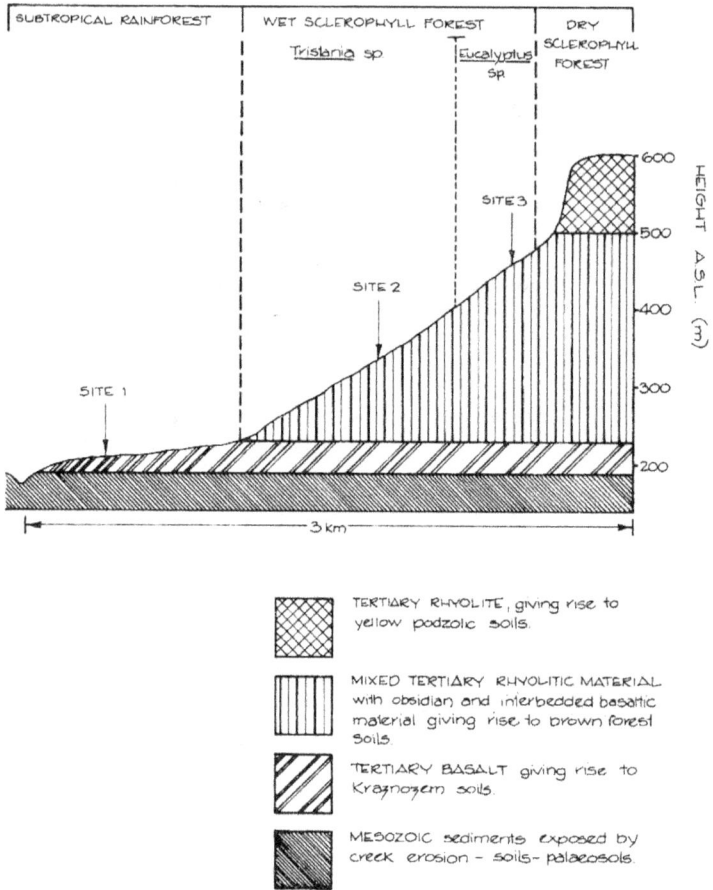

Fig. 9 Landscape profile at Terania Creek: Rainforest occurs on deep fertile soil in the most sheltered position. (Courtesy John Turner)

fertility declines and exposure increases upslope at Terania Creek, they give way to wet and then dry sclerophyll forests (Fig. 9). Rainforests grow on sites that were physically protected from Aboriginal fire. The rainforest at Terania Creek burnt only once in a millennium. Turner carbon dated a tree in the rainforest to about 900 years of age.[1] The rainforest regenerated immediately after the fire. It was quite capable of sustaining itself under the natural fire regime. Ross Florence thought that the Terania Creek rainforest developed "possibly as part of a successional sequence".[2]

There is neither evidence nor logic to support this. The rainforest grew back quickly after one fire in a millennium because it was perfectly suited to the site and it reinforced the natural, very infrequent, high-intensity wildfire regime. Palynology (the study of pollen in sediment cores) shows that natural fire refuges, where rainforests grow, expanded and contracted with climatic cycles.[3]

Mark Westoby (a prominent ecologist who was named NSW Scientist of the Year in 2014) was moved by rainforest 'campaigns' to declare to the Australian Ecological Society in 1984:

> We should stop wittering about ecosystems being fragile. I have yet to hear the media describe any ecosystem as anything but fragile. … The tropical forests are contracting in area not because they are susceptible to subtle ecological side-effects which our scientific knowledge has been unable to anticipate, but because huge amounts of labour, power and fossil energy are being systematically devoted to felling and removing the structuring species which make these forests what they are.[4]

In 1982 New South Wales' premier Neville Wran had firmly established his green credentials by ordering a halt to selective logging that retained the structure of rainforests. After Bob Carr returned Labor to power in 1995 he outdid Wran by 'protecting', in new national parks, all the extensive rainforests in New South Wales that had already been selectively logged.

The untenable theory of secondary succession and the wilderness mentality that prevails in Australian academia have created a mystique about rainforests and a fundamental flaw in conventional Australian ecology. Rainforests are 'worshipped' as the ultimate 'climax community'. Too many ecologists simply don't recognise that the pre-European distribution of rainforests was controlled by four interacting factors, and that changing a single factor will not allow rainforests to cover the landscape. The flawed theories have their genesis in the cool, moist, tall eucalypt forests of Tasmania where lightning strikes are uncommon and rarely start fires.[5]

Aboriginal influence on the Tasmanian landscape had already been reduced by a disease epidemic, probably influenza introduced by sailors or sealers, before European settlement. Protector of Aborigines, George Augustus Robinson noted in 1831:[6]

> from observations made during this journey combined with the testimony of the natives themselves, I ascertained beyond a doubt, that many of those districts which had been formerly peopled by the Aborigines are now unoccupied; the once resident tribes being utterly extinct, a fact which was evinced by the dense overgrown underwood.

Robinson rounded up and deported Aborigines who survived disease as well as battles against European invaders, and had avoided kidnap by Bass Strait sealers. The firestick was completely extinguished by 1833, only three decades after settlement. When Darwin climbed Mount Wellington in February 1836 he had a clear view over an extensive area because there was no smoke. By the time that ecologists such as Gilbert, Jackson and Mount started to develop hypotheses about the interaction of fire and vegetation, the natural patterns in the forests had mostly been obliterated.

Nevertheless, in 1965, forest ecologists Cunningham and Cremer were able to describe rainforest understorey in wet sclerophyll mountain ash forest and differentiate it from invasive "wet sclerophyll scrub" in moist forests of mountain ash or messmate:

> Rain-forest
>
> This is usually found in areas where annual rainfall exceeds 40 inches (1000 mm) and there have been no fires for over 100 years. It is characterised by the species *Nothofagus cunninghamii* ... and *Atherosperma moschata* ... but is often more complex and includes *Eucryphia lucida* ... *Anodopetalum higlandulosum* ... *Phyllocladus aspleniifolius* ... and several other minor species. The ferns, *Dicksonia antartica* ... *Blechnum procerum* ... *Histiopteris incise* and *Hypolepis rugulosa* may be present.

Wet Sclerophyll Scrub

This is the usual type of understorey found in the second growth stands less than 80 years old throughout the State. It is very common also in mature and overmature old growth stands where its age is always less than that of the overstorey. Its mature height ranges from 20 to 80 ft depending on species. The characteristic species are *Pomaderris apetala* ... *Olearia argophylla* ... *Phebalium squameum* ... *Bedfordia salicina*.[7]

These descriptions show that only sites with high rainfall and topographic protection from fire can support the full complement of species characterising true rainforest in Tasmania. A different suite of species requiring less rainfall or topographic protection can escape from moderately sheltered gullies and invade naturally open forests in the absence of frequent mild fires. The scrub understoreys are younger than the old growth forests because it took some time for them to climb up the slopes and spread through open forests. Nearly a century earlier, Howitt remarked that these same scrub understoreys had invaded the originally open forests of South Gippsland:

It may also be inferred, from the constant discoveries during the process of clearing of blackfellows' stone tomahawks, that much of this country, now covered by a dense scrub of gum saplings, Pomaderris apetala, Aster argophylla [now *Olearia argophylla*], and other arborescent shrubs, that the country was at that time mainly an open forest.[8]

It is unfortunate that Cunningham and Cremer referred to **wet sclerophyll** scrub. I will argue that rainforest subcanopies define true wet sclerophyll forest, whereas scrubs invade naturally open, moist and dry forests. In my book, wet sclerophyll forests are eucalypt forests that grow on very fertile and moist sites together with rainforest subcanopies. They occur as even-aged stands, regenerated by very infrequent, high-intensity, stand-destroying fires.

Several recent papers have examined age class structure and regeneration processes in supposed wet sclerophyll forests. Each has identified anomalies in stand structure or regeneration processes compared to implied but unstated definitions. The anomalies are: a preponderance of mixed-aged stands; and/or a lack of evidence that regeneration occurs as a result stand-destroying wildfires.[9] Many supposed wet sclerophyll stands are uneven-aged and have scrub understories. Such stands in North Queensland historically had open grassy understoreys, and recent changes in structure have been misrepresented as evidence of natural succession to rainforest.[10]

In 1968, Professor Jackson's famous hypothesis *Fire, Air, Water and Earth – An Elemental Ecology of Tasmania* immortalised the flawed theory of succession to rainforest. It did not give earth its due. Jackson stated that "There would appear to be no aspect, soil-type, or edaphic situation within this region which cannot be occupied by cold temperate rainforest".[11] Tony Mount's alternative hypothesis gave earth its due by acknowledging that vegetation is determined by both site and climate,[12] but it downplayed the human element by proposing that vegetation governs fire regime according to the rate at which fuel accumulates.[13]

Jackson recognised that human fire influenced vegetation directly by burning it and indirectly by its influence on soil fertility, but he suggested that "Ecological Drift" resulted from "chance occurrences of short or long intervals between two successive destructive fires". Though he initially thought that rainforest could grow anywhere in Tasmania but for fire, Jackson fine-tuned his hypothesis of ecological drift over three decades. He increasingly recognised the importance of earth (by way of topography and aspect), as well as the firestick, to an "Elemental Ecology". In 1999, Jackson gave a paper before the Royal Society of Tasmania detailing his refined hypothesis.[14]

He proposed that in the absence of the firestick, all high-rainfall areas (> 1200 mm) in Tasmania, except for exposed areas in the Alps, would be covered in rainforest. Medium rainfall areas (> 800 mm) would have eucalypt forest on northerly aspects and rainforest on southerly aspects because eucalypts are limited by light and rainforest trees are limited by water. There would be no mixed forests: "in the absence of frequent fires, the boundaries between the communities would be sharp and the present intergrading would not occur". Low-rainfall areas would carry eucalypt forest with pockets of rainforest in very sheltered gullies on southerly aspects, and Olearia or Bedfordia scrub in less-sheltered gullies.[15]

Jackson considered that heaths, sedgelands, grasslands and savannas had been created by Aboriginal burning through its direct impacts on vegetation as well as its impacts on soil fertility and drainage. He analysed climatic and palynological records from Tasmania and New Zealand showing that vegetation change during the last glaciation in Tasmania proceeded differently than it had during the previous three glaciations in Tasmania and during all four cycles in New Zealand. Twice as much open vegetation occurred in Tasmania under the influence of the firestick as would be expected based on climate alone.[16]

Peter McIntosh and colleagues built on Jackson's palaeological work and confirmed one aspect of his hypothesis whilst possibly challenging another. They showed that the firestick had created infertile poorly drained soils in Tasmania compared to New Zealand where it was absent. On sheltered southerly aspects in Tasmania, that were less frequently accessible to the firestick, soils are more fertile, better structured, and support taller eucalypt trees of different species than those on drier northerly aspects. But the understoreys typically contain Pomaderris and Olearia.[17] They are scrub understoreys, not rainforest understoreys.[18]

In Jackson's hypothesis, fires less than forty years apart will eliminate rainforest species and change "mixed forest" to "wet sclerophyll forest" whilst 350 years without fire will eliminate eucalypts and convert mixed forest to pure rainforest.[19] Like Cunningham and Cremer, Jackson differentiated scrub understoreys of Bedfordia and Olearia from rainforests of myrtle-beech (*Nothofagus cunninghamii*) with one or two other trees. However he didn't distinguish myrtle as a forest tree up to 40 metres high in cool-temperate rainforest[20] from myrtle as a dwarf tree in low-closed forest, otherwise known as scrub, on poor sites.

Tasmanian foresters noticed in the 1930s that stands of alpine ash started to die out after only fifty years without fire.[21] Some of the dying stands had sclerophyll shrub understoreys with occasional stems of myrtle, and others had almost pure understoreys of thick myrtle. This chronic decline of alpine ash was initially reported as succession to rainforest.[22] But the eucalypts were dying whether the invading understorey was sclerophyll shrubs or saplings of myrtle, and they were dying well before their natural lifespan. When the invading understoreys were felled, the eucalypts didn't recover, but when they were felled and burnt, the trees recovered and were soon growing nearly three times as fast as trees in untreated stands.[23] This chronic decline was obviously not natural succession to rainforest, and the understorey was not, in itself, the cause of decline. Changing soil properties in the absence of fire have since been found at the root of similar problems across Australia.[24]

In *Ecology and Silviculture of Eucalypt Forests*, Ross Florence proposed that wherever the climate is suitable for rainforest, vegetation 'progresses' along a gradient of soil fertility from heath through dry eucalypt forest, tall eucalypt forest, mixed eucalypt forest and rainforest, to pure rainforest.[25] He proposed that chronic decline of alpine ash is a rapid succession to rainforest because alpine ash is "close to its environmental limits" at high altitudes and post-European wildfires and logging have "destabilised natural processes" and "exposed the limits of the eucalypts' capacity to cope with" fertile soils".[26] In fact, as its name suggests, alpine ash is perfectly suited to high altitudes, and stands of wildfire and logging regrowth grow vigorously so long as fire is not excluded for too long.[27] Alpine ash suffers chronic decline when the environment is changed by disrupting the natural (pre-European) fire regime.[28] This is not natural succession to rainforest.

Florence considered that much of the chronically declining forest in coastal New South Wales is "wet sclerophyll" forest. He referred to an imaginary eucalypt-rainforest interface where eucalypts are at the limits of their tolerance of fertile soils, and boundaries between pure rainforest and pure sclerophyll forest shift as "secondary succession" occurs following disturbance by fire:

> conventional wisdom holds that … regular burning by Aboriginals created more clearly defined forest boundaries than are apparent today, and maintained a more open forest floor. An alternative view is that the sharper boundaries were related primarily to the effect a complete canopy has on soil conditions, i.e., a high rate of

water use, the incorporation into soil of a large amount of high C:N ratio litter and the negative influence this can have on the soil microflora, nutrient mineralisation and other soil biological processes.[29]

This alternative view is incorrect because eucalypt forests do not have a closed canopy[30] and eucalypt species that grow in wet forests thrive on the fertile soils. Grassy balds and open eucalypt woodlands commonly abutted pure rainforests before Aboriginal burning was disrupted.[31] It was the burning of the eucalypt litter and cured grasses at the rainforest boundary that kept mesic invaders out of the grassy eucalypt systems. This mineralised some nutrients and volatilised others, providing balanced nutrition for the grasses and eucalypts. Most importantly it prevented accumulation of nitrogen in the soil and killed woody seedlings.[32]

The occurrence of chronic eucalypt decline is a sure indication that a stand is not a true wet sclerophyll or mixed forest on a high-quality site that is capable of supporting rainforest in the long-term absence of fire. Chronic decline occurs on relatively dry, infertile and poorly structured soils created by Aboriginal burning.[33] These soils naturally support open, grassy forests and woodlands.[34] Florence considered that "the mixed sclerophyll forest-rainforest community may occur extensively without necessarily having contact with the rainforest proper".[35] But he mistook moist forests invaded by scrub understoreys[36] for true wet sclerophyll forests. Florence referred to supposed rainforests on infertile, poorly structured soils in Tasmania as follows: "A critical question to start with is why do such soils support rainforest at all? Unfortunately this question cannot be answered at present".[37] The answer is simply that they don't.

The alleged "dramatic expansion of rainforest" in Tasmania from the mid-1850s[38] was merely a case of scrub invasion after disturbance to the balance of nature. The apparent dilemma about rainforest on poor soils arises from a loose definition of rainforest. One or two mesic species escaping their natural refuges from fire do not constitute a rainforest. The supposed rainforests are depauperate scrubs dominated by myrtle. The scrubs can't develop into forests because the soils are not sufficiently fertile and well structured.

Similarly, there has been a problem with the definition of wet sclerophyll forests. Gilbert and Jackson both referred to eucalypts over rainforest as mixed forest.[39] Gilbert as well as Cunningham and Cremer differentiated between mixed forest and eucalypt forest with a tall shrub understorey including Pomaderris, Olearia and Acacia.[40] Jackson referred to "wet sclerophyll eucalypts" in "mixed forest".[41] Eminent Australian silviculturist Max Jacobs didn't refer to wet sclerophyll forests.[42] Florence's book doesn't define wet sclerophyll forest precisely, but it refers to "forests with tall-boled sclerophyll emergent over a stratum of rainforest elements".[43]

Florence used the terms "mixed forest", "wet sclerophyll", and "eucalypt-rainforest interface" interchangeably. He defined "tall open forest" as forest at least 30 m tall in which:

> there may be an understorey of ferns, tree-ferns, and small trees and shrubs which are mesophytic rather than xerophytic or sclerophytic in character. The term 'wet sclerophyll forest' has traditionally been used to describe this type ... where a canopy of tall sclerophyll-type trees dominates a secondary stratum of plants which may be largely of rainforest origin.[44]

He appears to have taken his lead from New South Wales' Silviculturist George Baur, who defined wet sclerophyll forest as: "Tall forest (over 100 feet in height), frequently with a scattered understorey of small trees, and with a mesomorphic shrub layer (which may however be destroyed by burning) and ground herbs".

In 1965 Baur defined dry sclerophyll forest as: "A single tree layer, with a frequently dense, sclerophyllous shrub layer and the ground stratum sparse. Usually under 120 feet in height". By 1989 he had revised the definition of dry eucalypt forest understorey: "with a sometimes dense, sclerophyllous shrub layer and the ground stratum often sparse unless burning has promoted grass at the expense of shrubs".[45]

Baur worked during the time that forest management transitioned from fire suppression to firestick forestry.[46] His definitions reflect the transition. The "mesomorphic shrub layer" in "wet sclerophyll" forest that can be "destroyed by burning" is invasive scrub. When foresters grasped the firestick in the mid-twentieth century they were able to restore open grassy conditions in moist forests that had been invaded by mesic scrub, and in dry forests that had been invaded by sclerophyll scrub. What Baur termed wet scleropyhll forest is more accurately described as moist shrubby forest.

Published classifications of Australian vegetation typically refer to 'Tall open forests' with trees greater than 30 m in height that have a variety of understorey types. For example the *Atlas of Australian ... Vegetation* notes of "Tall open forest" that:

> Most of these forests have a distinctive lower stratum of low trees and tall shrubs ... Fire-determined successions within the understoreys have been identified ... Rainforest species do occur in the understorey of some eastern forests, though in Tas. there are areas where the lower stratum is a closed forest dominated by Nothofagus cunninghamii.[47]

Thus tall eucalypt forests have been classified in one group whether the understorey is rainforest, soft-leaved shrubs, hard-leaved shrubs or open and grassy. Furthermore there is confusion in the literature about natural (pre-European) fire regimes and changes in forest understoreys with successive changes in fire regimes. Tall eucalypt forests have been officially designated by NPWS as wet sclerophyll forests in *Native vegetation of New South Wales and the ACT*:

> This combination of hard-leaved 'sclerophyllous' trees and soft-leaved 'mesophyllous' understorey plants sets the wet sclerophyll forests apart ... The wet sclerophyll forests, also known as tall open forests ... may be divided into two subgroups ... Compared with the shrubby subformation, the more open understorey of the grassy forests reflects their slightly drier habitats and their transitional status toward the grassy woodlands.[48]

This self-contradictory definition of wet sclerophyll forests is not useful. It reflects ignorance of Australia's ecological history, because the "grassy subformation" represents the natural state that was maintained by grazing and burning after Aboriginal management was disrupted, whereas the "shrubby subformation" is a consequence of shrub invasion after management by neglect.[49]

I suggest that the only way to finally resolve the persistent confusion about the nature and classification of tall eucalypt forests is by recognising wet sclerophyll forests with true rainforest subcanopies as distinct from moist eucalypt forests that are naturally grassy. Moist eucalypt forests are accessible to the firestick and are prone to decline and scrub invasion when 'protected' from fire because their trees, understoreys and soils were shaped by tens of thousands of years of Aboriginal burning.[50]

In examining "the rainforest-sclerophyll relationship", Florence[51] would better have discarded the unworkable concept of secondary succession and recognised that the firestick maintained sharp boundaries between rainforest and eucalypt forest, corresponding with changes in exposure, aspect, slope and/or soils. Chronic decline of eucalypts and scrub invasion can then be seen as consequences of true disturbance – the dousing of the firestick. Instead, Florence had an each way bet on the Jackson model based on climate, aspect and firestick, as well as the Mount model where site conditions and vegetation type set fire frequency, and the firestick doesn't count. His confusion was evident in the conclusion that "In all probability one or the other may be more relevant to a particular ecological situation, and there will be something of both in most situations".[52]

David Bowman apparently started his career as a disciple of Professor Jackson and progressed to colleague. Not surprisingly, his definition of rainforest refers to fire and encompasses scrubs as a type of rainforest:

> In Australia the term 'rainforest' has diverged from the original conception of luxuriant forests that develop in drought-free environments; indeed some Australian ecologists use the perplexing term 'dry rainforest'. This idiosyncratic terminology has arisen because the definition of Australian rainforests hinges not on moisture stress, but sensitivity to fire relative to the surrounding vegetation.[53]

Rainfall is certainly not a single factor that defines rainforest nor does it singly determine the moisture status of sites where rainforest grows. Rainforests and eucalypt woodlands or grasslands grew side by side under the same rainfall regime.[54] Accessibility to fire as influenced by temperature, topography and soil nutrient status are also important. But neither does sensitivity to fire define Australian rainforests. They are typically mixtures of trees with a range of relevant traits and responses to fire. For example, the leaves of the sassafras tree (*Atherosperma moschatum*) contain volatile compounds that promote crown fires, and many rainforest species, including sassafras and myrtle, are able to resprout after fires.[55]

Bowman acknowledged that "There is no consensus as to what constitutes rainforest in Australia". In evaluating the factors that determine where rainforest grows, he accepted whatever definition had been used in each study that he examined. So his conclusion that everything hinges on fire relies on including as rainforests, communities such as bottle tree scrubs[56] that grow in dry country and are not forests. Most people probably wouldn't see these as rainforests. I think that a simple definition of rainforest from the *Atlas of Australian … Vegetation* would go as close to a consensus position as any that I have seen. The Atlas defines rainforest as closed forest – trees at least ten metres tall with at least 70% foliage cover in the tallest stratum.[57]

I would go a bit further and suggest that people mostly think of rainforest in terms something like this: a mixed species forest with a closed canopy of mostly soft-leaved trees, growing under conditions of relatively high moisture availability on fertile, well structured soils, which is capable of sustaining itself indefinitely in the absence of disturbance. Remember that disturbance is a factor that is outside the range of experience of the ecosystem, as discussed in Chapter 5. The natural (pre-European) fire regime is not a disturbance.

Dry rainforest, as I know it in northern coastal New South Wales, grows on very sheltered, steep, easterly aspects but receives less rainfall than the higher ferrosol[XIX] plateaus above, which carry subtropical

---

XIX Fertile, well structured, soils high in iron, and without a clayey lower horizon that impedes drainage.

rainforest. The soils are moderately fertile, being enriched by erosion of soil from the higher plateaux, but not as well structured. Similar soils on drier, gentler, more exposed aspects support grassy eucalypt woodlands that are prone to decline in the absence of fire. The natural fire regime is probably infrequent fire that burns from grassy woodland into the low scrub stratum of the dry rainforest when leaves are dessicated by extreme drought. It doesn't fit the *Atlas* definition of rainforest because the leaves of the upper stratum don't cover 70% of the ground, and the lower stratum doesn't reach ten metres in height. I think that dry rainforest is a simple and descriptive name. If fire and/or grazing are excluded from adjoining eucalypt woodlands, some species from dry rainforest can invade the woodland. Dry rainforest cannot, because woodland sites are not sheltered enough and their soils are not good enough to permit development of emergent rainforest trees.

One or two rainforest species do not necessarily make a rainforest. Bowman used this fact to play down the role of Aboriginal colonists in creating the Australian landscape:[58]

> The archipelago of fire-sensitive fragments of rainforest in a sea of flammable vegetation has been interpreted as evidence of the tremendous impact of landscape burning by the Pleistocene colonists based on the assumption that the prehuman landscape was dominated by rainforest. This interpretation overlooks the remarkable diversity of the fire-adapted biota and does not address the question of how this ensemble of species developed.

He set up a 'straw man' by claiming that "The most popular explanation for these massive changes (the extinction of megafauna and contraction of rainforests) relates to the alleged impacts of the first Australians … Frequent burning triggered an evolutionary and ecological expansion of flammable plants". Bowman proceeded to burn the straw man by explaining that forty to sixty thousand years, or about 200 generations of eucalypts, was insufficient time for an evolutionary process to occur. He went on to summarise the aridification of Australia after Gondwanaland broke up and the continent drifted north, the penetration of tropical storms and lightning into the interior, and the antiquity of the northern eucalypts from the section Eudesmia. This explained how fire-dependent species evolved a long time before Aboriginal colonisation, but Bowman subsequently lost his way by arguing, against the palynological[XX] records, that fire-dependent species displaced the Gondwanan rainforest species "from all but the most fire-sheltered habitats" before human colonisation.[59]

Bowman claimed that "The classic example of this is the dynamic balance between the giant mountain ash *Eucalyptus regnans* and the southern beech *Nothofagus cunninghamii*". He cited Jackson's *Elemental Ecology* in support. But Jackson's data show no dynamism in the mixed forest. There is a consistent relationship over the life of the stand between the physically dominant eucalypts and the numerically dominant rainforest trees.[60] They grow together on sites that are physically, chemically and climatically capable of supporting true rainforests. Every few centuries, a lightning strike or spot fire during drought and extreme weather burns the mixed forest and regenerates both eucalypt and rainforest trees.[61] These true wet sclerophyll forests, together with true rainforest, grow on sheltered sites where they were confined by Aboriginal burning.

Palynology shows that rainforest species dominated humid zones during the last interglacial. They declined as sclerophylls increased with glaciation so that hard and soft leaved plants were 'running neck and neck'. After Aborigines arrived, hard plants surged ahead and soft plants retreated into their fire refuges.[62] There was no need to 'throw in' evolution and confuse the issue. It was really all about competition, and the firestick 'changed the game'. Wet sclerophyll forests are unique in that they are outside the reach of the firestick because they are protected by rainforest, and ignition occurs in exposed, elevated fuels above a rainforest sub-canopy. They are accessible to lightning fires or ember storms under extreme weather conditions during major droughts.

Jackson embraced the unworkable concept of succession and looked for a reason why Tasmania wasn't covered with rainforest. He eventually found it in the firestick.[63] But given the long presence of Aborigines, a

---

XX Palynology is the study of pollen grains obtained from sediments.

better question would be why there is any rainforest at all. From this perspective the earth receives due recognition as an element in the pattern. Bill Gammage's magnum opus restored Aborigines to their rightful place in the 'Grand Scheme', but it understandably downplayed the earth's role.[64] Too many ecologists have made the same mistake. Gammage argued that trees now encroaching grasslands show that grasslands were created by Aborigines. This is true, but he goes too far in suggesting that Aborigines created grasslands wherever they wished. He wrote of Kangaroo Valley: "Without fire this is climax rainforest country".[65] Actually most of it, like most of Australia, is physically incapable of growing rainforest.

Gammage considered that Kangaroo Valley's mosaic of vegetation was maintained by at least four distinct fire regimes. I'm sure this is true, but geological maps show that four distinct substrates helped to shape the pattern, along with topography, aspect and microclimate. Aborigines worked with these elements. They didn't burn against the grain as Gammage suggested: "This book shows that people often broke these parameters ... Climate, soil, altitude, aspect, nutrients ... they cannot explain all, or even most, 1788 landscapes".

These parameters, together with the firestick or grazing, can actually explain all landscapes in 1788 as well as in 1988. Grassy flats and slopes on the north bank of Kangaroo River and along the flanks of Barrengarry Creek are on heavier soil in a drier rainshadow area that is much frostier than the surrounding open forests. They grew either open red gum woodlands, or grasslands where the trees had all been burnt down. Grazing keeps the woodlands open by controlling woody seedlings. The sandy bed of Kangaroo River grows river oak (*Casuarina cunninghamiana*). This is not rainforest country.

'Climbing up' from Kangaroo Valley to the Southern Highlands, Gammage discussed a "thoughtful article" by Kevin Mills wherein:

> his Yarrawa Brush map shows both eucalypts and rainforest across his key soil boundaries, and of rainfall he concedes, 'The location of the eucalypt – rainforest boundary in the west cannot only be due to a decrease of rainfall'. ... I suggest that his well-researched

vegetation anomalies show people working with and against the country in 1788.

It is not so. I travelled through this country on my way to and from boarding school each term over five years, and many times on a motorcycle whilst attending University in Canberra. I can attest that the vegetation in the area is shaped by exposure to, or topographic shelter from, freezing westerly to southwesterly winds that howl across the tablelands and are funneled up the valley above Wingecarribee Swamp, nearly to Robertson. Rainforest occurred on sheltered northeasterly aspects to the north of Robertson and sheltered southerly aspects south of Robertson. Eucalypts occurred on exposed southwesterly aspects. Grasses and sedges dominated the exposed, boggy, frosty valley floor to the northwest of Robertson. A quick comparison of topography against Mills' map of the 'Yarrawa Brush'[66] should remove any doubt.

Forest encroaching grasslands since burning was disrupted illustrates that Aborigines moulded the vegetation, but the type of forest encroaching doesn't necessarily illustrate the clay that was used. Red gums are aggressively invading some grassy balds in the Bunya Mountains because there is still a source of seeds in the surrounding woodlands. They can't overrun balds surrounded by rainforest simply because there is no seed source.[67] Except for river red gum seeds dispersed by floods, eucalypt seeds germinate or die where they fall or are blown from the tree. There is no store of eucalypt seed in the soil. Gammage rightly points out that old roots remaining in the ground show that the balds had trees in the past. The roots are still there because they are durable eucalypt roots rather than 'softwood' roots of rainforest species.

Fensham and Fairfax found no association between grassy balds and environment, but they didn't construct a hypothesis and design an experiment to test it. Instead they jumped straight into data collection and statistical analysis.[68] The balds were made where eucalypt woodlands grew, and woodlands grew where fire could run freely. Their coarse categorisations of slope and aspect missed the subtle interactions underlying flammability. Even so, their data showed that there were nearly twice as many balds on dry, northerly and westerly aspects compared to more sheltered, easterly and southerly aspects. The larger balds were on steeper slopes[69] because fire runs faster and further up steep slopes.

Gammage's picture of a bald illustrates some of the subtleties. Rainforest grows on steep, sheltered, easterly slopes; grass and eucalypts grow on more exposed, gentle, upper slopes. He suggested that:

> It is not easy to burn rainforest to make grass, or to keep back eucalypts and other fire friendly species while protecting fire sensitive bunyas. To do all these at once evidences expert botanists and fire managers over many generations.[70]

But it's not really that complicated. Aborigines burnt grasslands and woodlands when hunting wallabies, and lit campfires in the hollow butts of widely spaced red gums whilst they gathered and feasted on bunya nuts. These fires prevented eucalypts from invading grasslands and converted some woodlands to grasslands when all the red gums eventually burnt down. Aboriginal fires trickled across grassy flats and down gentle upper slopes killing new seedlings. They raced up steep exposed slopes, but they couldn't run on steep sheltered slopes where rainforests lived. Ludwig Leichardt described a similar situation near the Hunter Valley in 1843:

> The base of both mountains is covered with forest, the principle tree being the black butt. The westerly flancs are covered with a rich grass, devoid of trees at Piri and with very few ones at Mt. Royal. The easterly flancs are covered with a rich brush to the even edge of both mountains.[71]

Fire management is much more complicated since the ascendancy of the wilderness mentality has created three dimensionally continuous fuels that will not burn under mild conditions, but explode into fire storms under severe conditions. Many ecologists and foresters, especially in Victoria and Tasmania, now regard these landscape bombs as 'wet sclerophyll' forests because the subtle environmental pattern enhanced by Aborigines has been obliterated by post-European homogenisation. Cunningham's and Cremer's work[72] helped to reveal this, but their term "wet sclerophyll scrub" confused the issue.

European settlement of Victoria commenced in the 1830s and extensive shrub encroachment had occurred by 1851. This fuelled the first megafire that blew up on Black Thursday, 6th February, and consumed five million hectares.[73] The scrub bounced back, and people born from that time on, came to regard scrubby forests as a natural phenomenon. A textbook on 'multiple stable states' published by Oxford University claimed that "When Europeans began settling southern Australia in the 1860s, they were confronted with vast expanses of wet-Eucalypt dominated jungle that extended eastward across the Gippsland plains".[74] But the explorer Strzelecki described the Gippsland Plains somewhat differently:

> The region eastward of the chain in the direction to Corner Inlet presents a totally different aspect. At the Latitude 37⁰, or about the sources of the river Thomson, the spurs are less ramified, and of considerable height and length, shaping the intermediate ground into beautiful slopes and valleys, which ultimately resolve into a fine open plain, richly watered, clothed with luxuriant grasses and fine timber, and offering charming sites for farms and country residences. Viewed from Mt. Gisborne, Gipps Land resembles a semi-lunar amphitheatre walled from N. E. to S. W. by lofty and picturesque mountain scenery, and open towards the S. E., where it faces, with its sloping area the uninterrupted horizon of the sea.[75]

The author of the textbook was referring to Strzelecki's description of a "broken inhospitable region" defined by two ranges that "by contrast enhance the beauty of Gipps Land".[76] I furnished him with a copy of Howitt's first-hand history of the Gippsland Plains describing how woody thickening occurred after Aboriginal burning was disrupted,[77] but I received no response.

Ron Hateley was a forester who thought that unnaturally scrubby eucalypt forests were wet sclerophyll forests. In 2010 he published *The Victorian Bush, its 'original and natural' condition*. Hateley stated that:

> The idea that almost every part of the continent would have burned in the same way ignores … wet sclerophyll forests where eucalypts like Mountain Ash (*Eucalyptus regnans*) and Alpine Ash (*Eucalyptus*

delegatensis) do not produce seed until 15 or 20 years of age. ...
At higher elevations aboriginals' impact on the environment was
probably minimal.[78]

These statements were misleading. Almost every part of the continent
was burnt frequently but not in the same way. Mountain ash and alpine
ash forests are mostly not wet sclerophyll forests, and it doesn't matter
how long trees in moist forests take to set seed because they are not killed
by mild or moderately intense fires. Surveyor Townsend observed that
Aborigines burnt right across the high mountains during their annual
feasting on bogong moths, and alpine ash was known as black gum[79]
because its rough bark was blackened by fire.

Hydrological work in mountain ash forests within Melbourne's water
supply catchments has shown how fire, air, earth and water interact to
determine the natural distribution of rainforest, wet sclerophyll forest
and scrubby eucalypt forest. Rob Vertessy and colleagues measured soils,
vegetation, precipitation, solar radiation, evapotranspiration and runoff
over a period of calibration, then they precisely and accurately predicted
runoff over a decade. Myrtle II was a small (32 ha) gauged catchment,
carrying old-growth mountain ash moist forest with an understorey of
acacia and olearia scrub. There was a small area of sassafrass – myrtle
beech rainforest and a small patch of wet sclerophyll mountain ash forest.
The catchment occupies a sheltered southerly aspect on deep, fertile, well
structured soils with about 1700mm rainfall.[80] Rainforest grows in the
bottom of the creek, extending up both main arms in a v shape and
joining across a steep southerly face not far above their junction. Wet
sclerophyll forest grows on a very slight rise, between the two arms and
below the steep southerly slope.[81]

The hydrological data show that rainforest grows in the most
sheltered positions with the largest upslope catchment area. Wet
sclerophyll mountain ash forest grows in an equally sheltered position
with a very small upslope catchment. Moist mountain ash forest grows
in less sheltered positions with a wide range of upslope catchment areas.
In the rainforest, the watertable is never more than about a half a metre
below the soil surface. In the wet sclerophyll patch, the water is always

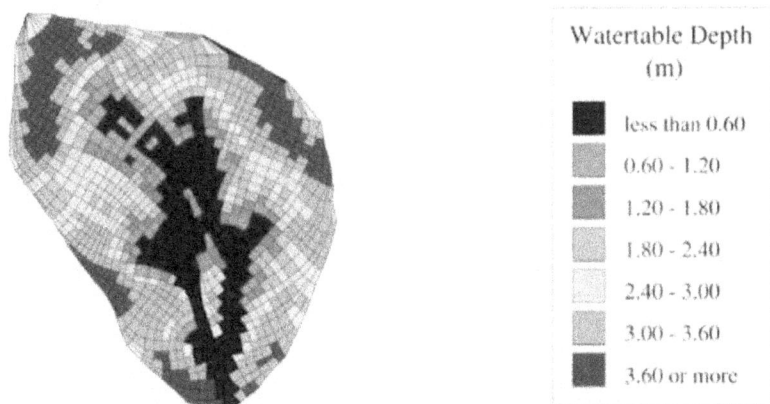

Fig. 10 Myrtle II Catchment: depth to watertable at a point in time. Lighter-shaded cells (deep watertable) carry wet sclerophyll forest at the bottom of the catchment and moist forest on midslopes. (Courtesy CSIRO)

more than a metre down. The depth to water in the moist ash forest varies seasonally according to aspect and slope. This pattern is apparent in Fig. 10 where the lighter-shaded cells in the bottom of the catchment occupied by wet sclerophyll forest are surrounded by the dark cells where rainforest grows. Lighter-shaded cells on midslopes are the habitat of moist mountain ash forest.

On lower slopes, the depth to water can vary seasonally between a half a metre and two and a half metres.[82] These are the areas where scrubby understoreys grew under Aboriginal management. Fires burning from the open ash forests occasionally penetrated these areas when they were dry. But only once every few centuries, did drought and extreme weather allow lightning or ember storm to ignite a fire in the wet sclerophyll forest. Then it burnt to the ground. Hard and soft plants germinated or sprouted, and the wet scleophyll patch grew back in the same place it had always occupied. The surrounding rainforest usually didn't burn because there was no dry fuel under the closed green canopy.

Gammage says that "Eucalypts topping rainforest indicate land people once went to great trouble, working against the country, to clear and keep clear".[83] This is not correct. Eucalypts overtopping real rainforest

– i.e., real wet sclerophyll forest – indicate land that was inaccessible to the firestick. Eucalypts overtopping scrub (the more common situation) indicate land that Aborigines maintained by frequent mild burning.

I was fortunate to have observed the subtle pattern of rainforest, very limited areas of wet sclerophyll forest, and expanses of open grassy forest before it was almost completely lost from northern New South Wales. Wet sclerophyll forests grow on sites that are physically and chemically capable of sustaining rainforests but are sufficiently exposed to sun and wind, or to drought, as to be flammable under severe conditions of climate and weather. In tropical and subtropical areas, these sites are refugia where a small number of eucalypt species managed to survive amongst expanding rainforests as climate ameliorated after the most recent glacial period.[84]

There was a good example to the southwest of Dome Mountain near Urbenville in Northern New South Wales. In the midst of a flat shelf on a ferrosol carrying subtropical rainforest there was a small knoll that supported a hectare of Sydney blue gum (*E. saligna*) forest with a well developed subtropical rainforest subcanopy. The blue gum were obviously several hundred years old and were producing large amounts of litter that would be flammable under drought conditions. This stand may have regenerated after a lightning strike several hundred years ago. But there are more extensive tall eucalypt forests on higher ridges about 500 metres to the north and about a kilometre to the west. At some time of drought and extreme weather during the past half a millennium these forests may have delivered embers to spark a fire that regenerated the blue gum knoll.

Blue gum is very closely related to flooded gum (*E. grandis*). Both are often referred to as wet sclerophyll species. For example, Baur wrote of blue gum "usually tall wet sclerophyll forest" and of flooded gum "tall wet sclerophyll forest … the type relies on catastrophic site disturbance (e.g., fire, cyclone) to regenerate and normally occurs in the gullies in small, even-aged stands intimately mixed with stands of rainforest".[85] But this is not the typical situation. Both species may occur as small patches of wet sclerophyll forest in slightly exposed or occasionally droughted positions within rainforest. More typically, they occur as naturally grassy stands that are now mostly invaded by scrub. Baur's type descriptions unfortunately 'hybridised' these distinct formations to come up with non-existent types. The error is evident within his description of Sydney Blue Gum Type: "Frequent fires produce a grassy understorey, while in the absence of fire an understorey of rainforest species develops in many, though not all sites".

A typical grassy stand of flooded gum was depicted in *Forest trees of Australia*, second edition 1962[86] (Fig. 11, p. 262). This stand of flooded or rose gum with pink bloodwood (*Corymbia intermedia*) surrounded a waterlogged depression carrying bangalow palms (*Archontophoenix cunninghamiana*). On another page (p. 59) there was a photograph of the same forest type on a similar site where fire had been deliberately excluded. This was captioned: "A rain forest scene". But there was no subcanopy of soft-leaved trees, only an understorey of tall shrubs and a few bangalow palms. Most modern observers would probably describe it as a 'wet sclerophyll forest'. However it is obvious that under Aboriginal management it was grassy forest because the trees are clearly older than the shrubs. The site is climatically suitable but physically and chemically incapable of supporting subtropical rainforest. Bangalow palms at both sites indicate impeded drainage and waterlogging. There are heavy clay subsoils created by thousands of years of frequent burning.[87] Lignotuberous species such as bloodwoods grow with the flooded gums. These species are unable to sustain themselves on true wet sclerophyll sites because lignotubers die out under prolonged shading.[88]

Doc Jacobs was arguably the 'father' of eucalypt silviculture. He referred to a small group of forest types as "very tall eucalypt forest on sites of high quality". Jacobs included flooded gum "in restricted sites only"[89] because he apparently recognised the dichotomy that I have described. Some of the open grassy stands that were originally maintained by Aboriginal burning, have since been logged, regenerated and 'protected' by foresters so that they are denser, taller and much shrubbier. But very tall (> 70 m) flooded gums from restricted high-quality wet sclerophyll sites in tropical Queensland are genetically uniform

Fig. 11 Grassy forest of flooded gum and pink bloodwood. Bangalow palms grow in a waterlogged depression behind. (Courtesy Forestry and Timber Bureau)

compared to their genetically more diverse relatives in naturally grassy forests (e.g., Fig. 11) further south.[90] Obviously they were isolated in the refuges where they persisted whilst rainforests expanded after the last glacial.[91] I have no doubt that the restricted and isolated wet sclerophyll stands in southern Queensland and northern New South Wales would also show relatively uniform genetics compared to the more widespread, naturally grassy stands.

Wet sclerophyll flooded gum was recently portrayed in the scientific literature as a pioneer rainforest species.[92] Given its ecological history and genetic uniformity it is clear that it is a relic rather than a pioneer. What is currently regarded by many ecologists as rainforest expansion or natural succession in the wet tropics is really scrub invasion of naturally grassy forests that were created and maintained by Aboriginal burning. A few soft-leaved species are invading vast areas of eucalypt woodlands and grassy forests in the absence of frequent mild burning.[93] But firestick, air, earth and water define the natural habitat of rainforests. Some species that are naturally confined to rainforest or deep sheltered gullies by the firestick can escape that confinement when it is doused. Rainforests cannot, because fire is not the only factor that confines them.

Pittosporum undulatum was, in nature, a component of some small pockets of rainforest on some very sheltered, or very rocky near-coastal sites in southeastern Australia.[94] It is now a major weed of urban bushland around Brisbane, Sydney, Melbourne, Hobart and Adelaide.[95] Pittosporum forms a closed scrub of up to 8,000 stems per hectare in old bushland reserves north of Sydney Harbour[96] where the natural vegetation as described by Tench and Mitchell was barren woodland. It is running rampant together with phytophthora on the foreshores of Sydney Harbour where eucalypts are dying.[97] Adelaide is well outside its natural range as are the Azores where pittosporum is the major woody weed. There is a proposal over there to use it as the feedstock for a sustainable bioenergy industry.[98] Nowhere, inside or outside its natural range, is invasion by pittosporum regarded as expansion of rainforest. But it is exactly the same process that is occurring with other native rainforest species in Tasmania and northeastern Queensland. The invasion of woodlands by red gum or cypress scrubs is another facet of the same process, and the demolition by Australian governments of sustainable industries dealing with the problem (see Chapter 11) contrasts starkly with the pragmatic proposals to deal with an aggressive exotic invader in the Azores.

William J. Bond and various colleagues have produced a considerable volume of literature on this global problem of scrub invasion and loss of biodiversity in the absence of human fire. For example:

> for ecosystems with a long history of fire, there is concern over the cascading consequences of anthropogenic fire suppression. In tall grass prairies, and comparable grasslands elsewhere, fire suppression has led to the loss of as many as 50% of the plant species ... Small herbaceous plants with high light requirements for growth and seedling establishment are the worst affected. Changes in faunal composition have also been reported, for example, in dry dipterocarp woodlands, where fire suppression has resulted in a marked loss of termite species ... Even greater species losses occur where fire suppression leads to complete biome switches, such as from savannas to forests ... there are also reports of complete biome switches from grassy ecosystems to closed forest and thicket. Brook & Bowman (2006) reported a landscape-wide expansion of forest (42% increase) in the Australian monsoon tropics over the past five decades, while Russell-Smith et al. (2004) reported a near doubling of forest patches in savannas in northeast Queensland from 1943 to 1991. In Kansas prairies, gallery forest increased in area by 69% from 1939 to 2002, fragmenting the prairie landscape (Briggs et al. 2005). In South Africa, scrub forest has invaded savannas in conservation, commercial ranching and communally farmed areas, with increases from 12% to 68%, 5% to 55%, and 3% to 27%, respectively, from 1937 to 2000 (B. Wigley, W.J. Bond & T. Hoffmann unpublished). ... Biome switches have cascading consequences for biodiversity and ecosystem function.[99]

Nearly 20% of four million hectares of coastal woodlands and forests in NSW are switching to scrub.[100] In the wet tropics of North

Queensland, more than half the area of the major eucalypt types has already been lost to scrub.[101]

Australian Rainforests – Islands of green in a land of fire is Bowman's explanation of how our rainforests came to be where they are. He argued that the pattern was established by lightning fires for millions of years before human arrival. But he contrarily suggested that:

> The frequency of fire in the landscape is of critical importance in controlling the distribution of rainforest. ... Frequent fires prevent the maturation of seedlings or re-sprouts, and ultimately eliminate rainforest from a landscape except where topography creates fire refugia. ... At the time of European colonisation, Aborigines carefully managed landscapes with fire and the extent of rainforest was influenced by their burning practices. Following European colonisation, the cessation of Aboriginal landscape burning has led to the expansion of rainforest in some places and its contraction in others. ... If we let nature take her course, we may see the ultimate dominance of fire-adapted plants at the expense of the ancestral and less fire-adapted rainforests.[102]

I say 'you can't have your cake and eat it'. It is clear that the interaction of climate, soils and topography shaped vegetation before human arrival. Since human arrival, vegetation has been shaped by man using the firestick in combination with climate, soils and topography. Man is a part of nature. To let nature take her course we need to rekindle the firestick. Then we will see an end to forest decline and scrub invasion. Rainforests, wet sclerophyll forests and natural scrubs will once again be Islands of dark green in a sea of grassy ecosystems.

Endnotes

1 Turner 1984.
2 Florence 1996.
3 Singh *et al.* 1981.
4 Westoby 1984.
5 Gilbert 1959, Jackson 1999.

6 Flood 2006.

7 Cunningham and Cremer 1965.

8 Howitt 1891.

9 Hickey *et al.* 1999, Turner *et al.* 2009, Stanton *et al.* 2014b.

10 Stanton *et al.* 2014b.

11 Jackson 1968.

12 Florence 1996, p. 62.

13 Mount 1964.

14 Jackson 1999.

15 Ibid.

16 Ibid.

17 McIntosh *et al.* 2005.

18 Cunningham and Cremer 1965.

19 Jackson 1999.

20 Jackson 1968, Fig. 1.

21 Ellis 1964.

22 Ellis *et al.* 1980.

23 Ibid.

24 Ellis and Pennington 1992, Jurskis 2005, Turner *et al.* 2008, Close *et al.* 2011, Horton *et al.* 2013.

25 op. cit.

26 Florence 1996, p. 64; 2005, p. 265.

27 Ellis *et al.* 1980.

28 Ellis and Pennington 1992, Jurskis 2005, Turner *et al.* 2008, Close *et al.* 2011, Horton *et al.* 2013.

29 Florence 2005.

30 Gilbert 1959, p. 133; AUSLIG 1990.

31 Leichardt 1843, Gammage 2011.

32 Turner *et al.* 2008, Jurskis *et al.* 2011.

33 McIntosh *et al.* 2005.

34 Jurskis 2005, 2011b; Turner et al. 2008; Jurskis *et al.* 2011.

35 Florence 2005.

36 Cunningham and Cremer 1965.

37 Florence 1996, p. 63.

38 Ibid., p. 64.

39 op. cit.

40 op. cit.

41 op. cit.

42 Jacobs 1955.

43 Florence 1996, p. 20.

44 Florence 1996, 2005.

45 Baur 1965, 1979; Anon. 1989.

46 Pyne 2006, p. 2.

47 AUSLIG 1990.

48 Keith 2004.

49 Jurskis 2002.

50 Jurskis 2005, McIntosh *et al.* 2005, Turner *et al.* 2008, Jurskis *et al.* 2011.

51 op. cit.

52 op. cit., p. 62.

53 Bowman 2003.

54 Leichardt 1843, Gammage 2011.

55 Boland *et al.* 1984.

56 Bowman 2000.

57 op. cit.

58 Bowman 2003.

59 Ibid., p. 7.

60 Jackson 1968, Fig. 1.

61 Jackson 1968, 1999.

62 Jackson 1999, Kershaw *et al.* 2002.

63 op. cit.

64 Gammage 2011.

65 Ibid., p. 232.

66 Mills 1988.

67 Fensham and Faifax 2006.

68 Fensham and Fairfax 1996.

69 Ibid.

70 Gammage 2011, p. 71-2.

71 Leichardt 1843.

72 op. cit.

73 Howitt 1891, Adams and Attiwill 2011.

74 Petraitis 2013, p. 2.

75 Strzelecki 1845.

76 Ibid., p. 64.

77 op. cit.

78 Hateley 2010.

79 Townsend 1846.

80 Vertessy *et al.* 1993.

81 Langford and O'Shaughnessy 1977.

82 Vertessy *et al.* 1998.

83 op. cit.

84 Singh *et al.* 1981, Fig. 4.

85 op. cit.

86 Forestry and Timber Bureau 1962, p. 19.

87 McIntosh *et al.* 2005.

88 Jacobs 1955.

89 Ibid.

90 Jones *et al.* 2006.

91 op. cit.

92 Tng *et al.* 2012.

93 Stanton *et al.* 2014a.

94 Floyd 1989.

95 Rose 1997, Low 2003.

96 Rose 1997.

97 Cahill *et al.* 2008.

98 Lourenço *et al.* 2011.

99 Bond and Keeley 2005, Bond 2008.

100 Jurskis and Walmsley 2011.

101 Stanton *et al.* 2014a.

102 op. cit.

12a

HOW ABORIGINES MOST LIKELY
MADE AUSTRALIA

On ne saurait faire une omelette sans casser des oeufs

Robespierre

It seems to me that Bill Gammage's book subtitled *How Aborigines made Australia*,[1] is really about how Aborigines maintained Australia for a very long time after their ancestors made it. To get an idea of how the original settlers made Australia we need an idea of what they had to work with. Matthew Flinders provided some small insights by describing the vegetation, fire regimes and soil on some uninhabited islands of wilderness off the southern coast of the continent.[2] These had dense, shrubby, even-aged regrowth forests with huge loads of timber felled by firestorms after lightning strikes.

On Kangaroo Island the soil was:

> judged to be much superior to any before seen ... upon the south coast of the continent ... The depth of the soil was not particularly ascertained; but from the thickness of the wood it cannot be very shallow. Some sand is mixed with the vegetable earth, but not in any great proportion; and I thought the soil superior to some of the land cultivated at Port Jackson [Sydney].[3]

Unfortunately the only available pollen and charcoal record on the Island, from Lashmar's Lagoon, doesn't go back far enough to show us what species of trees and bushes were there and exactly what happened when Aborigines first arrived.[4]

However the pollen and charcoal record from Lynch's Crater on the Atherton Tableland near Cairns in northern Queensland extends back to the end of the penultimate glacial period,[5] and clearly shows the different impacts of climate change and Aboriginal burning on Australian vegetation. Up until the end of the previous interglacial, most of the pollen blowing or washing into the crater was from rainforest plants. About 85% of the rainforest pollen was from soft-leaved flowering plants. Hard-leaved plants were contributing less than 10% of the total pollen. Small amounts of charcoal were washing in periodically. As the climate became colder and drier, sclerophyll pollen increased to about half of the total and coniferous pollen increased to about two thirds of the rainforest pollen. Charcoal increased substantially and was deposited continuously for a few thousand years before dropping back to a similar pattern as before.

Immediately after human arrival during a long period of relatively stable climate, 42,000 years ago, there was a dramatic peak in charcoal delivery to the crater.[6] This declined over thousands of years but remained continuous at higher levels than before. Pollen of rainforest trees also declined to a few percent of the total. Eucalypt and casuarina pollen doubled compared to previous levels.

When it started to warm and wet up again, after the glacial maximum about 20,000 years ago, charcoal once again increased dramatically. But pollen composition didn't change until the end of the glacial period about 11,000 years ago. Then there was another peak in charcoal production as rainforest pollen began to increase substantially. After this, charcoal declined dramatically along with sclerophyll pollen until about 6,000 years ago. By this time eucalypt and casuarina pollen, as well as charcoal, were back to levels similar to the previous interglacial.

It is apparent from these records that Aboriginal fire initially produced much charcoal as it converted mixed forest into sclerophyll forest. Charcoal production then declined in the open grassy sclerophyll woodland, even as it became drier with the approach of the glacial maximum. Charcoal increased once again as the climate warmed and rainforest expanded, because Aborigines were still able to burn the thickening, mixed forests. By the end of the glacial period, rainforest prevailed again because most of the Atherton Tableland was too wet to burn. Eucalypt woodland persisted only in areas receiving less than around 1300 millimetres annual rainfall. Isolated areas of wet sclerophyll forest persisted in exposed positions within wetter areas[7] because they were accessible to lightning ignitions or to embers coming from nearby open forests, under extreme weather conditions during occasional droughts.

Returning to Kangaroo Island, we can examine the changes in vegetation and fire regime after the demise of the Aboriginal population about five thousand years ago. Under Aboriginal management, there were open woodlands of drooping sheoak (*Casuarina stricta*) and scattered acacias and eucalypts with sedges and ferns in the groundlayer. For roughly the first millennium after the Island was cut off from the mainland, there was a constant low supply of charcoal to Lashmar's Lagoon. During the succeeding millennium or so, there were small peaks and troughs in charcoal supply as the declining or possibly itinerant population of Aborigines became unable to manage the land intensively.

In the centuries after the last known date of Aboriginal occupation, there was a sharp spike in charcoal production, and in wattle pollen. Also there were sudden declines in pollen from grasses, ferns and sedges, and a more gradual increase in eucalypt pollen. Singh and colleagues argued that the spike in charcoal and dramatic changes in vegetation 4800 years ago were probably due to a marginal drying of the climate. However the only evidence they proffered was the reduction in spores from ferns and in pollen from sedges, so the argument is rather circular. About 2500 years ago, there was a dramatic increase in charcoal production and sharp decrease in wattle pollen back to levels similar to the Aboriginal phase.

I think it is likely that the first extensive high-intensity wildfires triggered massive germination of a large store of acacia seeds that had built up in the soil as Aboriginal management declined. The wattle scrub quickly choked out the groundcovers, leaving bare ground. It is possible that a similar process created the brigalow scrubs in central Queensland after Aboriginal populations were decimated by smallpox. On Kangaroo Island, eucalypts slowly expanded outwards from isolated trees and edges as younger recruits successively reached maturity. As the eucalypts coalesced into forest they suppressed the wattle scrub. Prolific growth of cypress scrub in western New South Wales after European settlers disrupted Aboriginal burning[8] was a somewhat similar process, but eucalypts cannot establish under the dense canopy of cypress scrub.

Singh and colleagues deduced that a huge increase in charcoal deposition from 2500 years ago indicated a regime of infrequent high-intensity wildfires, and Flinders' observations support this deduction. The dominance of eucalypt pollen and the low levels of acacia suggest that a eucalypt forest had established, producing large quantities of fuel that retained some moisture under the shade of the canopy which also suppressed wattles. When lightning strikes coincided with dry fuels under extreme weather conditions, firestorms erupted.

I think that Aborigines made Australia by applying the firestick wherever and whenever they could. Trees and shrubs that are killed by fire were gradually eliminated or pushed back into physical refuges. Mature, fire-resistant woody plants and grasses were promoted at the expense of susceptible species and woody seedlings. As canopy declined and grasses increased burning became easier. Open, sunny and breezy areas with cured grasses were easily burnt with mild fires. Resprouting shrubs were set back as soon as grasses cured, and eventually eliminated by repeated fires. A new balance was established. Fire frequency was governed by accessibility to the firestick. Aborigines didn't burn rainforests because they were inaccessible to the firestick. John Turner's work at Terania Creek showed that combustible fuel occurred in subtropical rainforest only once in a millenium.[9] Aborigines worked with climate cycles that were discernible in human timeframes. They were unable to halt the advance of rainforest on the Atherton Tableland when climate ameliorated at the onset of the current interglacial. The charcoal and pollen records show that they tried and failed when it became too wet.[10]

Gammage remarked many times about the juxtaposition of fire-sensitive and fire-dependent plants in the landscape, stating that some woodland trees such as cypress are sensitive to fire.[11] But these trees are not sensitive to mild fire in their natural habitat of open grassy woodland (see Front and Back Cover). On the other hand, dense scrubs are inaccessible to the firestick but accessible to firestorms which kill mature trees. Gammage wrote: "In western New South Wales Jim Noble found that scrub took over grassland within twenty years of settlement. All the problem seedlings, White Cypress, Budda, Yarran, Bimble Box, were fire sensitive". But **all** new seedlings are fire-sensitive. Bimble box is lignotuberous, so seedlings are only sensitive to fire for a short time until a lignotuber develops. All these trees are insensitive to mild fires running through cured grasses.

Fire-sensitive species are those that cannot survive moderate fires as established plants. Frequent burns in grassy woodlands are low-intensity fires. Almost 90% of budda and yarran survived devastating wildfires in the Western Division of New South Wales during the summer of 1974-75. Survival of cypress was less than 20% but most of the mortality was of small trees in dense clumps.[12]

After bushfires once again devastated the Western Division of New South Wales in 1985, claiming five lives and 40,000 head of stock, Cobar Forestry District was reopened. I was appointed to assess the cypress timber resource across more than a million hectares of leasehold land. The intention was to salvage commercial timber from trees killed in the fires. As it turned out, salvage was unnecessary because sawlog-sized trees grew mainly in open forests and woodlands that didn't suffer crown fires. Relatively few commercial trees were killed and there weren't any substantial areas of dead, log-sized trees to salvage. Gammage suggested that Aborigines were very skillful in maintaining "fire sensitive" cypress trees,[13] but many of the oldest and healthiest cypress trees remaining in the landscape today have survived regular and intense stubble burns by wheat farmers.[14]

Scrubby mountain ash forests are widely regarded as fire-sensitive 'wet sclerophyll' forests needing stand-replacing fires to persist in the long term. Gammage wrote:

> Mountain Ash has thin bark and almost any fire kills it, yet its leaves are the most flammable of all eucalypts, and it drops more litter than most. It recovers from seed, so needs fire to clear the ground and lay ash. If a fire comes it makes sure the fire is big and hot, for if no fire comes in 400 years or so it dies, and scrub chokes its

seedlings. ... Whole forests rage with a ferocity none can fight and few survive. Almost certainly people managed Mountain Ash in winter: they lived on the coast in summer. Perhaps for generations they winter burnt to keep edges back and clearings open, sensibly beginning when a forest was young. Yet a winter cool-burn could not ensure enough heat for those wet forests to kill and replace themselves. That needs a dry summer, yet it was done. It must have been a time for brave men, their families safely on the coast.[15]

Harking back to Myrtle II catchment (see Chapter 12), it is clear that moist mountain ash forest can't burn in winter. Receiving 1700 millimetres annual rainfall with a winter/spring maximum, on a southerly aspect which hardly sees any sun in winter, it is never dry enough.[16] Aborigines fired mountain ash in summer. Mature, wide-spaced trees have a thick 'stocking' of rough bark up to fifteen metres above the ground[17] that insulates them from the heat of fires. There were 20 trees per hectare, each more than a metre in diameter in Myrtle II.[18] Even young regrowth forests with scrubby understoreys can withstand moderate fires in dry fuels under mild weather. Some stands were unscathed when they burnt at night during the Black Saturday fires in 2009, which merely burnt their understoreys. On the other hand, stand replacement fires in true wet sclerophyll forests are uncontrollable and deadly. Aborigines had neither the need, nor the desire, nor the capability to attempt them.

Flinders description of wilderness on Kangaroo Island gives us another clue how Aborigines made Australia:

> After all the researches now made in the island, it appeared that the kangaroos were much more numerous at our first landing place, near Kangaroo Head, than elsewhere in the neighbourhood. That part of the island was clearer of wood than most others; and there were some small plats which seemed to be particularly attractive, and were kept very bare. Not less than thirty emus or cassowaries were seen at different times.[19]

Naturally, Aborigines newly arrived in Australia would firstly seek out such plats[XXI] where there was abundant game, just as Flinders did on Kangaroo Island. They would, of necessity, extend their explorations through the sparsest woods first, and they would, of necessity, burn from the sunniest edges as they went, lighting whatever could be lit. Aborigines would have been drawn to travel through areas made accessible by lightning fires, and they would have maintained access by burning these areas again as soon as it was physically possible. They likely travelled across country to investigate smokes seen or smelt from a distance. Flood speculated that Aborigines were attracted by smoke, seen across 100 miles of sea, to originally investigate Australia.[20] Two separate large peaks in charcoal deposition around the time that Aborigines originally occupied Australia[21] provide some small support for this speculation. Megafires in pre-human Australia may have informed legends of land to the south that encouraged explorations by ancestral Australians.

But Gammage's explanation of how Aborigines made Australia is back to front:

> To convert eucalypts to grass people had to let fuel build up so fires could run, but burn often enough to kill seedlings, and maintain this over many generations until the old trees died. Burning most eucalypts every 2 – 4 years would in time make grassland, while burning a little less often would let some saplings survive and create open woodland. Both were common in 1788, some where trees and scrub grow thickly now, others kept clear for so long that they have lost their seed stock and re-tree only by edge invasion, but re-tree they do.[22]

Aborigines had to start from edges where there was cured grass, sunshine and wind. There was plenty of fuel in the mixed forests but it was inaccessible to controlled fire until Aborigines working from the edges gradually reduced the shade and increased the air circulation so that green and woody fuels dried out. They must have worked outwards, gradually increasing sunshine, wind and grass, and creating a shifting green, brown and black mosaic in their wake. But even burning every year or two wouldn't necessarily produce a grassland. This is evident

---

XXI Flats, plots or lawns.

from the very substantial areas of woodland produced by Aboriginal burning over many millennia. Incidentally, "edge invasion" is the only way that eucalypts (other than river red gum whose seeds are carried by floods) can encroach on grasslands because there is no "seed stock" in the soil. Seeds either germinate or die where they are blown or fall from the tree.

Grasslands were maintained or created by frequent burning on relatively difficult sites for woody seedling establishment (exposed, frosty, waterlogged, gilgai soils etc.) where abundance of resources attracted very intensive use by Aborigines, even if only periodically. For example, the grassy balds of the Bunya Mountains were maintained or created where Aborigines hunted wallabies using fire, and where large groups camped and feasted and burnt and trampled seedlings, and burnt camp fires in hollow butts of red gum or against fallen trees. Soils under grassy balds and under forests in these mountains are reported to be "intrinsically similar" but there are "usually black soils" under balds and "usually dark red soils" under forests.[23] Foresters know from experience that it is much easier to establish tree seedlings on the red soils than the black which are prone to both cracking and waterlogging.

Except for eucalypt lignotubers, Aboriginal burning killed most woody seedlings most of the time, whatever the soil. Occasional seedlings escaped fire when they developed in a temporary fire refuge such as bed of ash or charcoal where a log had burnt away or perhaps in the stump hole from a burnt-down tree. Aboriginal fires burnt litter and dry herbage. Aborigines used fallen timber for firewood and construction. Once they had established the supremacy of sclerophylls and grasses, there was no need for clearing fires or clean-up fires as conceived by Gammage.[24] Referring to cypress he wrote: "These examples suggest clean-up fires every 30 years or so, plus cool burns every 1-3 years to kill seedlings. Clearly great care went into managing so ready a regenerator".

But mild fires every year or so burnt any dry fallen timber that hadn't been used for construction or firewood at the same time as they killed young woody seedlings. There was nothing to "clean-up", and in any case the concept of hot clearing fires doesn't work. All Australian plants are ready regenerators after fire, otherwise they would not be here.

Some, like cypress and river red gum, are regular and prolific seeders, so frequent mild fires are needed to kill seedlings and maintain the health of established trees. Most eucalypts produce few but very persistent lignotuberous seedlings. Many trees and bushes, including many rainforest trees, resprout when killed by fire. Repeated hot fires don't clean-up country, they dirty it by stimulating germination of soil-stored seeds of wattles and other hard-seeded plants, stimulating release of aerially-stored seeds, and causing many woody plants to resprout from branches, trunk or base. Fuel accumulation is much more rapid after fierce fires than after mild fires that burn only litter and cured grasses.[25]

Gammage provided quotes alluding to this problem:[26]

'One went burning in the hottest and driest weather in January and February, so that the fire would be as fierce as possible and thus make a clean burn … the long-followed practice of regularly burning the bush in the hot part of the year has resulted in a great increase of scrub in all timbered areas except the box country. The fires forced the trees and scrub to seed and coppice, and in time an almost impenetrable forest arose.'

He added:

Yet in Gippsland no fire also results in dense scrub. At Venus Bay, Robinson noted, 'This tribe once powerful are defunct and the country in consequence is unburnt having no native inhabitants. This is the reason the country is so scrubby.' To make Gippsland 'park-like' and 'clean-bottomed' it must be burnt correctly, not 'as fierce as possible'.

The apparent paradox arises because Gammage confused the original work by which Ancestral Aborigines made Australia over a millennium or two, with the ongoing work of maintenance over forty or fifty millennia. Woodlands were open and grassy, as made by Aborigines, when Europeans took them over. Forested country had already started reverting to scrub when Europeans looked to expand their grazing enterprises into rougher country. It required rebuilding rather than maintenance.

Noongars in Western Australia kept jarrah open by burning in summer every three or four years.[27] Fires hot enough to scar mature jarrah trees occurred only once in a century[28] probably according to medium-long term climatic cycles. Forests have more shade and less grass than woodlands so they can't be burnt easily every year. Once the forest was opened up by ancestral Aborigines it could be easily maintained by burning. After each burn, it might take three years before enough litter and dry grass accumulated to carry running fire.[29] Then, when the fuel was dry in summer, a few Aborigines could light an edge from a woodland or a recently burnt patch of forest, and let it run until it reached another edge. The original making of open forests required more intensive effort and more time, working gradually from outside edges into the heart of the forest.

Modern attempts to control 'woody weeds' in the Western Division of New South Wales, provide some clues about the making of open country. Gammage mistakenly referred to some of this work as follows: "In semi-arid country two fires every five years are needed to clear hopbush, but it became a major pasture menace after 1788".[30] These attempts actually highlighted the problem that burning scrub infrequently makes it thicker in the long run through resprouting and seedling establishment. James Noble reported that:

> Because of the [temporary] reduction in hopbush competition, much less rain would have been required to generate sufficient fuel to carry a second fire. A second fire within five years would not only create a more open stand but would also eliminate hopbush seedlings before they became sexually mature and replenished the soil seedbank. By depleting seed reserves in this way, it is highly unlikely that further treatment would be required for a considerable time, probably thirty years or so.[31]

Further maintenance wouldn't be required for several decades in this case because grazing by domestic stock in the newly open and grassy vegetation would limit seedling recruitment.

When Aborigines were initially making Australia they had to burn again as soon as there was accessible fuel, whether they were following up lightning fires or their own edge fires. Two fires in close succession would be the absolute minimum requirement to open up country. Once open, maintenance was easy because seasonally cured fuel would then accumulate rather than evergreen trees and bushes, and shade.

Fire studies in dry forests at Eden showed that groundlayer plants and biodiversity declined within four years after fire as large shrub species increased in stature and density, shading out delicate species.[32] Aborigines made Australia by killing fire-sensitive trees and bushes, and then killing resprouts and woody seedlings as soon as they possibly could. This wasn't a science developed over millennia, it was a simple consequence of producing green pick, discovering or flushing game and maintaining access as they opportunistically followed the line of least resistance in expanding their economy outwards from natural grass plats and fresh water. After lightning fires, they would have burnt litter as soon as there was enough to carry fire, and grass as soon as it was cured. Woody seedlings were killed and mature trees and bushes survived unless they were sensitive to fire. Trees and bushes that were sensitive to mild fires disappeared from most of the landscape. Some of them survived in refuges that were inaccessible to the firestick. These now make up our rainforests and natural scrubs.

Flinders described such a scrub at the southern end of Keppel Bay on the Queensland Coast:

> I left the ship again in the morning, and went up the southern arm to a little hill on its western shore; hoping to gain from thence a better knowledge of the various streams which intersect the low land on the south side of the bay … Our attempts to reach the top were fruitless. It would perhaps have been easier to climb up the trees, and scramble from one to another upon the vines, than to have penetrated through the intricate network in the darkness underneath.

> Disappointed in my principal object, and unable to do anything in the boat, which could then not approach the shore within two hundred yards, I sought to walk upwards and ascertain the communication between the south and south-west arms; but after

much fatigue amongst the mangroves and muddy swamps, very little information could be gained.

He met up with large parties of up to twenty-five Aborigines on the shores of the Bay:

> It is scarcely necessary to say, that these people are almost black, and go entirely naked, since none of any other colour, or regularly wearing clothes, have been seen in any part of Terra Australis. About their fireplaces were usually scattered the shells of large crabs, the bones of turtle, and the remains of a parsnip-like root, apparently of fern; and once the bones of a porpoise were found; besides these, they doubtless procure fish, and wild ducks were seen in their possession.[33]

Aborigines used Keppel Bay intensively but they didn't use South Hill because it had nothing to offer them. They couldn't introduce fire into this mini wilderness because it was too damp and shady. There was no edge to work from because it was surrounded by mangroves and swamps. Flinders' botanists had a look at the scrub on South Hill while he was updating his charts, because rare plants are mostly found in scrubs.[34] The Hill "afforded a variety of plants; but they found little that had not before fallen under their observation".[35]

Wallaby Creek near Tooloom had grasslands, grassy woodlands and forests, dry rainforests and subtropical rainforests. Each had its place. Calaby found that there were more species of mammals in greater numbers than anywhere else in Australia.[36] Subtropical rainforest grows on a plateau with a deep fertile ferrosol soil and high rainfall. Below the plateau, a steep easterly slope and sheltered gullies had dry rainforest, and gentler, more exposed slopes had grassy eucalypt forest or woodland. Wallaby Creek is at the base of the slopes. East of the Creek is a flat where grassland and woodland grew. Upslope of the flat, on the westerly aspect, was woodland. North of the flat, as the valley constricted towards the head of the creek, dry grassy forest changed to moist grassy forest with taller straighter trees. An abundance of macropods gave the place its European name. Gammage would describe it as a rich "template".[37]

When I worked there in the early 1980s it was easily maintained by graziers and foresters using mild fires. The eucalypt forest was burnt every few years.

About a decade later, burning was deliberately reduced to 'protect' the environment, and the eucalypt forest and woodland was invaded by aggressive shrubs and saplings. Macropods lost some of their best forage as herbs and grasses were choked out. The soil and microclimate changed and eucalypts became sick.[38] Psyllids, bellbirds, koalas and mistletoes flourished. Rainforest boundaries didn't change because they occurred where the firestick makes no difference and never did. Earth, air and water defined the territory of the firestick. It was vast but it was not all encompassing.

The original taming of Australia was a difficult and drawn out process. Aborigines gradually turned mixed forests into sclerophyll forests and woodlands, and they extended grasslands through woodlands. Megafauna were extinguished along with some fire-sensitive plants. Other fire-sensitive plants were confined to refuges that were inaccessible to the firestick. Once made, the maintenance of Australia was easy. But when that maintenance lapsed the consequences were dramatic.

Gammage's view of the maintenance of the *Biggest Estate on Earth* seems to me, overly romantic:

> Fire let people select where plants grew. They knew which plants to burn, when, how often and how hot. This demanded not one fire regime but many, differing in timing, intensity and duration. No natural regime could sustain such intricate balances. We may wonder how people in 1788 managed this, but they clearly did. The nature of Australia made fire a management tool. No doubt some plants suffered during the learning centuries … plant communities embracing different fire responses thrived in 1788. Multiply this by Australia's 25,000 species, and a management regime of breathtaking complexity emerges. …
>
> Grass-seed eaters burnt after harvest, usually in late summer, relying on timing and damp plant bases to fetter the flames; tuber eaters mostly in early summer once tubers matured. Wet country

was lit whenever fuel was dry enough, which meant mostly in summer. ...

In the Centre old-growth fire sensitive plants are rare and in poor country, because in good country fuel builds up and fire kills them. Here was a dilemma (fire kills sensitive species but no fire lets fuel build up) and a solution (locate such species on poor land where less fuel means fewer fires). ...

When to burn grass might hinge on its varying growth from year to year, or on associated tubers or annuals, some killed by fire, others needing it to flower seed or compete. These needs are easily unbalanced. ...

Situating plants located grazing animals and their predators as precisely as a farmer with paddocks. 'The burning of country was not at all random', Kimber observed, there was greater attention to areas favoured by certain nutritious or otherwise useful plants and to areas favoured by certain animals ...

Understandably, fire was work for senior people, usually men. They were responsible for any fire, even a campfire, lit on land in their care. They decided which land would be burnt, when, and how, but in deciding obeyed strict protocols with ancestors, neighbours and specialist managers.

Graziers still burn off annually, though more for cattle than diversity as in 1788.[39]

I think that this view understates Aborigines' economic motivations, and downplays women's role in Aboriginal economies. It seems also to undervalue European pastoralists' attachment to land. Gammage's "tsunami of evidence"[40] makes it clear that burning by Aborigines was economically motivated and culturally entrenched. But this was also the case with High Country graziers, as they have demonstrated in their ongoing battle with green bureaucracy over more than half a century. It is still the case with many New South Wales' north coast graziers, such as those that maintain their part of the formerly rich template at Wallaby Creek while green bureaucrats allow scrub to choke out the public forests.[41]

Aboriginal male elders didn't have a monopoly on burning. Mitchell wrote: "In summer, the burning of long grass also discloses vermin, birds' nests, &c., on which the females and children, who chiefly burn the grass, feed".[42] Martu women are playing a substantial role in returning the firestick to the Western Deserts.[43] There is no doubt that Aboriginal burning was mostly very deliberate. It was also opportunistic according to fuel and weather conditions. Some was spur of the moment, such as in unexpected encounters, and some was accidental as noted by Howitt.[44] Aborigines achieved the possible. They couldn't do some of the things that Gammage suggested, such as managing stand-replacing fires in wet sclerophyll forests.[45] But they certainly burnt very purposefully and efficiently through constant practice.

Of "A clearing near Mudgeegonga", Gammage wrote:

> This air photo shows a grass plain lying in eucalypt forest, drained by a small creek. About 60 by 200 metres, it slopes gently west to where the creek drops over a 5-metre cliff and down a rocky gully into farmland. The landowner says the plain was 'always' there. Sapling stumps ring its rim but not its centre, indicating land kept open first by fire then by axe. The same soil continues into bordering box and stringybark, which cover steep enclosing hills. This is a brilliantly placed trap. Wallabies panicked on the plain would flee downslope and crash over the cliff, and survivors would be ambushed in the narrow gully.[46]

Here Aborigines worked both deliberately and opportunistically. The firestick was used together with earth, air and water to place the trap. The grass plat is on a flat cliff top. Obviously it is growing on skeletal soil over bedrock. Being in the drainage line, it alternates between waterlogging and drought. Being exposed to the west and enclosed by steep hills it is an oven in summer and a frost hollow in winter. It is a natural grassland with water, where wallabies congregate. It has food and water for Aborigines. Naturally they burnt it to attract wallabies. They couldn't extend it because the enclosing slopes are too steep and rocky to grow good grass, and trees have the natural advantage.

Sapling stumps around the rim indicate that trees can occasionally establish in favourable seasons. Whether they could ever grow to maturity is unknown. Aborigines may have extended the grass plat marginally in width or length. Perhaps there was originally a narrow strip of woodland. If so, evidence would be expected of the remains of gnarled roots engaged in fissures in the bedrock. No matter whether Aborigines created a hectare of grassland from grassy woodland, or marginally extended an existing grassland, they used commonsense and worked with the landscape. As "Kimber observed", they concentrated their efforts when and where there were most resources. They spent most time and effort where there was most to be gained.

There was most to be gained where there was grass together with water, on plains, in valleys, on shelves under steep slopes. There were many examples in the upper Clarence and Richmond valleys where I worked in the early eighties, including grassy balds next to creeks, lagoons and waterfalls. These areas had been used intensively by Aborigines, and there were still strong local communities. The mosaics of grasslands, woodlands, eucalypt forests and rainforests were then used by graziers who also maintained the template with fire. There was still a rich and varied fauna as noted by Calaby.[47]

Maintaining the rich template was not "a management regime of breathtaking complexity". Gammage wrote that "plants need particular and distinct fire at the right time and with the right frequency and intensity … Most curious, these different fires made similar plant patterns across Australia".[48] But it's not really curious that firestick, climate, soils and topography created recurrent patterns. Nor is it curious that seedeaters didn't burn until after the harvests; nor that tubers with green tops didn't get burnt. Aborigines burnt when and where there was fuel. That is – flammable organic matter. The time and place were determined by climate, topography, soil and length of time since the last visit of the firestick. Some plants and animals were favoured, some were extinguished, and some were confined to refuges that were inaccessible to human fires. Where the firestick was doused, plants and animals that had been favoured by it declined. Twenty four mammals were extinguished in NSW's Western Division after European settlement.[49] Trees and shrubs that had been kept in check by the firestick choked out grassy ecosystems.[50] Plants that had been confined to physical refuges invaded grassy areas. But plants that were confined by climate, topography or soils weren't released, and rainforests didn't expand.

Gammage was ambivalent about the environmental impacts of the firestick:

> Life and land may not have been perfectly balanced in 1788. … Fire in particular exacted a price … though probably not under 1788 management. … It reduces surface water, kills fire sensitive plants and animals, and impoverishes soil as compost becomes uncommon and ash the readiest nutrient. … Yet if people did jar the land's regenerative capacity, imbalances were slight and their threat remote. What came after 1788 was much more serious. Damage then was ignorant rather than willful, but revolutionary.[51]

In my view, life and land were perfectly balanced before Europeans settled in Australia. The firestick had changed the game about forty or fifty thousand years earlier. In 1788 it maintained a balance. After 1788, whitefellas disrupted the balance. It was revolutionary, but no more so than what had occurred when Aborigines changed the face of the continent and extinguished the megafauna.

Gammage argued that:

> More historical evidence exists than most scientists realise, but as well they and I differ in interpreting it. I value it; most of them devalue it … A habit for natural explanation emerges in scientists who decree against 1788 fire. Some deny it completely, some minimise its impact, some accept it but deny purpose or control. These deniers push their assumptions hard. …
>
> Before offering specific examples, I list … common assumptions and my responses. … 'The impact of 1788 fire was slight. Climate, soil, altitude, aspect, nutrients or growth habits suffice to explain Australia's vegetation'. … This book shows that people often broke these parameters; they cannot explain all, or even most, 1788 landscapes.[52]

Together with many of my colleagues who have experience in managing land using fire, I don't fit neatly with any of his listed categories or assumptions. I value the historical evidence greatly, but I differ from him in interpreting it. I have a bent for simple explanations of natural systems including man. I certainly don't decree against 1788 fire, I accept it with its great purpose and control. I argue that the firestick was critically important but had little environmental impact in 1788 because it was fundamental to the balance of nature. I certainly don't say that "Climate, soil, altitude, aspect, nutrients or growth habits suffice to explain Australia's vegetation".

I see 1788 fire as equally important with climate, soil, altitude, aspect and topography. These parameters together with the firestick explain all 1788 landscapes and all 2015 landscapes. Aboriginal people didn't break the other parameters using fire in 1788. Their ancestors broke the parameters that existed forty or fifty thousand years ago, before human fire reached Australia. They changed Australia's flora and fauna, and established a new balance.

In 1788, evidence of how Aborigines made Australia was scant. Mitchell saw some in the fossils of megafauna; in the absence of dingoes from Tasmania and of tigers and devils from the mainland; and in the woody thickening and loss of herbage and kangaroos after the firestick was extinguished.[53] He, like Flinders, Curr, Howitt and so many others, recognised the power of the firestick. None of them had the comprehensive evidence we have now about prehistoric changes in flora and fauna with climate change and with human arrival. Bowman downplays the latter[54] whilst Gammage explores neither. Ironically they agree on little other than that fire has the overwhelming influence on Australian vegetation. Bowman suggests that rainforests will disappear or shrink if we don't burn carefully.[55] Gammage wrote of Kangaroo Valley "Without fire this is climax rainforest country",[56] implying that rainforests would cover the humid zone of eastern Australia if we didn't burn. Neither view is supported by palaeology. Nevertheless both authors provide incontrovertible historical evidence that the firestick is the key to conservation of biodiversity in Australia.

Endnotes

1 Gammage 2011.

2 Flinders 1814.

3 Ibid.

4 Singh *et al.* 1981.

5 Ibid.

6 Singh *et al.* 1981, Rule *et al.* 2012.

7 Singh *et al.* 1981, Fig. 4.

8 Rolls 1981, Jurskis 2009.

9 Turner 1984.

10 op. cit.

11 op. cit.

12 Noble 1997; pp. 44-5, Plate 10.

13 op. cit.

14 Lacey 1973.

15 op. cit., p. 166.

16 Vertessy *et al.* 1993.

17 Boland *et al.* 1984.

18 Vertessy *et al.* 1993.

19 op. cit.

20 Flood 2006.

21 Mooney *et al.* 2011, Fig. 2.

22 op. cit.

23 Fensham and Fairfax 1996.

24 op. cit.

25 Birk and Bridges 1989, Jurskis and Underwood 2013.

26 op. cit.

27 Ward *et al.* 2001, Abbott 2003.

28 Burrows *et al.* 1995.

29 Ward *et al.* 2001, Abbott 2003.

30 op. cit.

31 op. cit.

32 Penman *et al.* 2008, 2009.

33 op. cit., pp. 129-31.

34 Oxley 1820, Jurskis 2009.

35 op. cit.

36 Calaby 1966.

37 op. cit.

38 Jurskis *et al.* 2003.

39 op. cit.

40 op. cit.

41 Horton 2012, p. 12.

42 Mitchell 1848.

43 Walsh 2013.

44 Howitt 1891.

45 op. cit., p. 166.

46 op. cit., p. 73.

47 op. cit.

48 op. cit.

49 Lunney 2001.

50 Noble 1997.

51 op. cit.

52 op. cit.

53 Mitchell 1839.

54 Bowman 2000, 2003.

55 Bowman 2000, p. 288.

56 op. cit., p. 232.

13

THE BIG JUMP

*Mere hunters, who absolutely cultivated nothing – the spear, the net, the tomahawk – could have produced no appreciable effect on the natural products of a large continent. Nor did they ; but there was another instrument in the hands of these savages which must be credited with results which it would be difficult to over-estimate. I refer to the **fire-stick***

Edward Curr

As school children in the 1960s, it was impressed upon us by our teachers that Darwin's advance over Lamarck was his recognition that traits acquired during the life of an organism were not inherited by its offspring. Our teachers illustrated the supposed differences in the theories by referring to Lamarck's idea that giraffes acquired long necks through constant stretching to browse trees. Imagine my surprise recently, upon reading *The Origin of Species*, to find that this was a furphy. Darwin wrote: "But we learn from the study of our domestic productions that the disuse of parts leads to their reduced size; and that the result is inherited". His theory clearly and explicitly incorporated inheritance of features modified by use or disuse.[1]

Thinking that maybe my memory failed me or I'd misinterpreted what I'd been taught, I 'surfed the net' and found that some academic institutions and even the American Association for the Advancement of Science are still perpetuating this misrepresentation of Darwinism. We did, however, learn at school of Gregor Mendel's contribution to genetics by experimental breeding of peas, though not that his work was initially rejected or ignored by the scientific community and rediscovered after his death. At university we studied genetics by experimental breeding of fruit flies and blowflies because they breed more quickly than peas. Forty years later, I was astounded to learn that a German scientist,

August Weismann, in the late-nineteenth century, rejected the notions of Lamarck and Darwin, and identified the mechanism of inheritance through "Germ-Plasm".[2] His outstanding contribution to science was not mentioned in our genetics unit or textbook at university.

Darwin's great advance was recognition of the struggle for existence and survival of the fittest. This recognition, along with the ancient principle that he often repeated – *Natura non facit saltum* (nature doesn't make jumps) – underpins my understanding of evolution or succession, competition, the balance of nature, disturbance and firestick ecology. That the ancient principle is not quite right is evident in Darwin's repeated references to "monstrosities" but Mendel's and Weismann's contributions to genetics allow us to accept it as a useful generalisation in ecology.

Darwin wrote that: "If we forget for a moment that each species tends to increase inordinately, and that some check is always in action, yet seldom perceived by us, the whole economy of nature will be utterly obscured". Tom White, in *The Inadequate Environment*, suggested that Darwin had it back to front:

> This view has it that the struggle for existence in nature follows remorselessly from the capacity of organisms to increase their numbers exponentially. But rather the reverse is true. The capacity of all organisms to increase their numbers exponentially follows remorselessly from the struggle for existence. ... The capacity to increase exponentially ... evolved because, only in populations in which females produced many offspring did sufficient individuals gain access to sufficient resources, and survive in each generation, to enable the population to persist. ... Because this is a finite world and in a finite world there was one sure consequence of the evolution of the first self-replicating entities. Sooner or later a resource, essential to their growth and replication, ran out. ... No longer could all replicate ... natural selection began to operate.[3]

It took me a long time to work out what White really intended to convey by this circuitous argument. Firstly he argued that only species with great fecundity could persist because resources were limited and there were few survivors in each generation. Secondly he seemed to argue that resource limitation was a consequence of fecundity. But resources are limited because the world is finite. Organisms compete against their own species and against other species for food. In order to survive and reproduce, they also have to 'evade' predators. Both competition and predation are forces of natural selection.

Darwin had argued in the following terms:

> The amount of food for each species of course gives the extreme limit to which each can increase; but very frequently it is not the obtaining food, but the serving as prey to other animals, which determines the average numbers of a species. ... Climate plays an important part in determining the average number of a species, and periodical seasons of extreme cold or drought seem to be the most effective of all checks.

White rejected the concept that predators regulate their prey over about 400 closely argued pages. He pointed out that predators are limited by their prey, that average numbers have no meaning in nature, and that there is no equilibrium density of a population – "only the maximum number that can survive each generation in a population that is pressing hard against the variable but limited supply of resources in its environment".[4]

But White agreed with Darwin that climatic fluctuations drive population fluctuations. In fact White could be regarded as an early 'climate sceptic'. He was so impressed when Sinclair and colleagues demonstrated a strong correlation between sunspot cycles and irruptions of snowshoe hares in Alaska that he added a note "in proof" to his book because it supported his prediction that further research could "indicate that variation in solar activity is the single most important factor influencing changes in the weather [and fluctuations in populations of plants and animals] on Earth". However, I digress. There should be no doubt that predators are limited by their prey, just as arbivores are limited by their trees. There should also be no doubt that predation and arbivory are evolutionary forces.

Arbivores and predators evolve their attacks as trees and animals that eat plants evolve their defenses. Unfortunately White dismissed this concept of a balance of nature. He claimed that "The tacit assumption in the balance of nature is that all species of organisms tend to produce an abundance of progeny that would survive to reproduce and lead to ever increasing numbers unless controlled by negative feedback mechanisms".[5] However, I make no such assumption. To me, the struggle for existence, White's "universal hunger for nitrogen" and the balance of nature are one and the same. They explain why *natura non facit saltum*. Evolution, as Darwin emphasised, is an exceedingly slow process. Predators cannot suddenly get a jump on prey because the individuals that they can't catch are the ones that pass on their inheritance. This is the balance of nature.

Darwin expressed it as follows:

> Species ... change much more slowly [than domesticated varieties] ... all the inhabitants of the same country being already so well adapted to each other, that new places in the polity of nature do not occur until after long intervals, due to the occurrence of physical changes of some kind.[6]

In *Life on Earth*, David Attenborough referred to palaeolithic cave paintings, and argued that "we are the only creatures to have painted representational pictures and it is this talent which led to the developments which ultimately transformed the life of mankind". To make the argument he adopted the wilderness mentality and downplayed the achievements of those renowned painters, the Australian Aborigines:

> The Aborigines ... live in harmony with the natural world around them, altering it not at all and making do with what it immediately provides. ... Cut off from our libraries [descended from cave paintings] and all they represent, and marooned on a desert island, any one of us would be quickly reduced to the life of a hunter gatherer.[7]

During the 1960s when we were being misinformed about Darwinism, Australian schoolchildren were also shown films depicting traditional

Aboriginal society in the Western Deserts. These films mostly portrayed Aborigines as noble savages in much the same way that Attenborough does. Having since read the early European accounts of Aboriginal economy and technology, I feel cheated. I can scarcely comprehend the psychological damage perpetrated on the descendants of this outstanding culture by those of the wilderness mentality. The traditional manufacture of musical instruments, tools, hooks, nets, poisons and fire; the ability to source water from natural wells on stony hills or from the roots of trees; possession of a mental GPS/GIS; the ability to survive comfortably whilst Europeans perished; elaborate burials, paintings, songs and dances; the totems, taboos and dreamings are equal in my book to any European galleries or libraries.

Curr described Aboriginal knowledge of country in the following terms:

> A remarkable faculty of the Blacks, which seemed nearer akin to instinct than to reason, is that by which he finds his way about the bush. ... It is also noticeable that whilst a white man's steering in the bush – directed by compass, or its substitute, the sun – is generally attended with some mental effort, that the Black finds his way without reference to anything of the sort, though he has names for all the cardinal points, and almost without reflection.[8]

Mitchell vividly described the well structured development of a "corrobory" and equated its art with the "delights of poetry". He repeatedly acknowledged the supremacy of Aboriginal culture over European culture in supporting his explorations.[9] Attenborough would have done well to wonder how the cave-painters could have worked without the firestick for illumination. In failing to ponder the transformation of Australia by hunter-gatherers he overlooked the one big jump in the story of *Life on Earth*. It is the firestick, not the pen, that sets man apart because the firestick gave him the capacity to influence the struggle for existence – to alter the balance of nature.

In *The Last Lost World*, Lydia and Stephen Pyne eloquently described the big jump in the following terms:

Controlled fire brought light and heat, but direct application most likely commenced with cooking … Heating potential foodstuffs improves their biological payoff dramatically … Simply by cooking food early hominims expanded the range and richness of their diet … There was a change in human morphology that reflected a change in diet … The most auspicious reforms … focus on the skull … No longer required to chew tough matter, the teeth and jaw can decrease, and the musculature needed to power them become less massive. That means the skull need not act as an anchor post to hold them; it can lighten and bulge upward without dietary and selective cost. It can accommodate a much larger brain.

The upshot was the largest leap in anatomical change recorded among the hominims: the most dramatic change in tooth size in six million years; the largest boost in body size; the hugest increase in cranial size (42%); and a sudden capacity to trek over whole continents. Something happened to endow Homo erectus with capabilities that restructured anatomy, physiology, and behavior in ways not easily deduced from environmental upheaval alone. Since then, while cranial capacity has increased, no other reconstruction of morphology has occurred on a comparable scale.

… The oldest hearths, evidence for fire tending, date to 1.4 million and 1.6 million years ago. The oldest archaeological evidence for fire making is 790,000 years ago in the Dead Sea rift valley. By then the basis for fire technology was fully realised, and granted the conservatism of Acheulean technology [primitive toolmaking] generally, it is probable that erectines had it from the beginning.[10]

Tom White does not appreciate the power of the firestick. To try and resolve arguments about regulation versus limitation of populations, he provides ample evidence that the world is green because herbivores are limited by their food.[11] Coevolution ensures that herbs slowly adapt their defences as herbivores slowly adapt their attacks. White is convinced that soil and climate control vegetative growth, and soil is fixed, so that herbivores can only proliferate when exceptionally good seasons increase the growth of herbage, or climatic stress reduces plants' defences and/ or improves their nutritive value. Thus he is unable to appreciate the dramatic consequences of Aboriginal arrival and then of European arrival. White confused variations in the balance between trees and arbivores according to climatic variability with the unnatural imbalance of chronic tree decline.[12]

South African ecologist William Bond studied savannas and recognised the importance of fire. But he devalued it by equating fire with browsing animals such as elephants.[13] A debate at cross purposes ensued with White focused on climatic limitation of plant growth, and Bond focused on global anomalies in the distribution of forests compared to the climatic potential for tree growth.[14] Consideration of Australia's prehistory and history can resolve this debate. Before human arrival, climate limited the regional distribution of both fire and forest. Aborigines upset the balance of nature and changed the regional distribution of fire and forest independently of climate. The firestick is much more powerful than a herbivore. It is not limited by climate. Man can use it in conjunction with short-term variations in weather to escape climatic limitations, because flammability as well as biomass governs fuel availability. The firestick can affect weather, soil formation processes, and regional response of vegetation to climate change.[15]

Edward Curr appreciated this in the mid-nineteenth century:

I am disposed to attribute to them [Aborigines] many important features of nature here ; for instance, the baked, calcined, indurated condition of the ground so common to many parts of the continent, the remarkable absence of mould which should have resulted from the accumulation of decayed vegetation, the comparative unproductiveness of our soils, the character of our vegetation and its scantiness, the retention within bounds of insect life (notably of the locust, grasshopper, caterpillar, ant and moth), a most important function, and the comparative scarcity of insectivorous birds and birds of prey. ... such constant and extensive conflagrations must also have had an influence on the thermometrical range, and probably affected the rainfall and atmospheric and electrical conditions.[16]

The power of the firestick is quite evident from the cataclysms that occurred firstly when Aborigines brought it to Australia, and secondly when it was doused by European settlers. Thirdly, Australia is currently suffering socioeconomic and environmental chaos because greens and academics have stolen the firestick from pragmatic scientific managers, and they don't know how to use it.

Endnotes

1 Darwin 1860, p. 474.
2 Weismann 1893.
3 White 1993.
4 Ibid.
5 Ibid.
6 op. cit., p. 321.
7 Attenborough 1979.
8 Curr 1883.
9 Mitchell 1839, 1848.
10 Lydia and Stephen Pyne 2012.
11 op. cit.
12 White 1986.
13 Bond and Keeley 2005.
14 White 2006, Bond 2006.
15 Jackson 1999, McIntosh *et al.* 2005, Walsh 2013.
16 op. cit.

13a

OF PRICKLY PEAR AND OTHER STUFF

As natural selection acts by competition, it adapts and improves the inhabitants of each country only in relation to their co-inhabitants

Charles Darwin

Next to the firestick, and domestication and cultivation of animals and plants, man's greatest impact on the balance of nature has possibly been the introduction of exotic species. In Australia, smallpox decimated Aboriginal populations soon after it was introduced to the north coast by Indomalaysian trepangers.[1] Following European settlement, rabbit plagues denuded large areas for half a century whilst prickly pear choked out tens of millions of hectares of farmland and pastures. The Enemy Release Hypothesis (ERH) is often advanced to explain the success of exotic species. ERH assumes that populations are regulated by their predators.[2] But we have seen that this is not correct.

Darwin recognised that exotics were favoured by natives' lack of evolutionary familiarity with them: "we need feel no surprise at the species of any one country, although on the ordinary view supposed to have been created and specially adapted for that country, being beaten and supplanted by the naturalised productions from another land".[3] Natives haven't been naturally selected to compete with or evade exotic species. There is no need for an ERH to explain the success of exotics. Their introduction is a disturbance – a new experience to the native ecosystem, which upsets the natural balance. However the success of biological control has often been seen as a corollary that supports ERH. How can it work unless predators do in fact control their prey?

The simple explanation is that exotic species initially encounter virtually unlimited resources because native species haven't evolved to compete with them or evade them. They, in turn, provide a virtually unlimited resource for their natural predators when introduced as biocontrol agents. The biocontrol agents increase exponentially until they 'hit the wall' of resource limitation. They cannot eliminate their food because they are limited by it. Foxes didn't eliminate rabbits and cactoblastis didn't eliminate prickly pears. Dingoes didn't hunt tigers and devils to extinction on the mainland, they out-competed them. Tigers and devils no longer had enough to eat because dingoes beat them to the slower and weaker animals that they had relied upon for food.

But smallpox and Aborigines didn't evolve together, which begs the question – Why were Aborigines not eliminated? Smallpox didn't eliminate Aborigines because it evolved rapidly. The fittest form was the least virulent – it killed its host slowly and therefore reproduced most successfully through the Aboriginal population. Darwin noted that natural selection is an exceedingly slow process occurring at geological timescales as environments change. Rapid evolution seems a contradiction in terms. But studies of successful exotics show that they often evolve rapidly in their new environment.[4] Sixteen of twenty three exotic plants studied in Australia had changed significantly in form within about a century of their introduction.[5] Being superior competitors, exotics increase exponentially into very large populations with much variability upon which natural selection can act when they hit the wall of resource limitation.

Rabbits are renowned for breeding rapidly. When they were introduced to virtually unlimited resources in the form of Australian plants lacking effective defenses against them, they certainly did. Australian rabbits quickly became different to their introduced ancestors. When myxomatosis was spread by mosquitoes in the warm, wet summer at the end of 1950 it caused 99% mortality of rabbits[6] because it was essentially exotic to Australian rabbits. However viruses breed even more rapidly than rabbits. Myxoma rapidly evolved into a less virulent form because this favoured transmission and reproduction.

Professor Frank Fenner, who gave his name to the School of Environment and Society at ANU, has sometimes been credited with biocontrol of rabbits in Australia and global control of smallpox. In fact he merely studied the evolution of myxomatosis from a highly virulent disease that decimated plagues of rabbits to a cohabiting organism in balance with reduced populations of rabbits. When an epidemic of human encephalitis coincided with the myxoma epidemic in rabbits, Fenner and two colleagues were goaded by a public health official to inject themselves with myxoma so as to prove that it wasn't causing human encephalitis. They apparently did this with no ill effects. Later he was prominent in the bureaucracy that eventually **certified** the eradication of smallpox.[7] Smallpox was finally eradicated by vaccinations because they eliminated its food – susceptible people.

Fenner was honoured as a Member of the British Empire because he helped to reduce Australian casualties from malaria in New Guinea during the Second World War. Later he worked as a microbiologist at the John Curtin School of Medical Research before becoming the bureaucrat in charge of that organisation. He was the founding Director of the Centre for Resource and Environmental Studies (CRES) at ANU in 1973. Costin joined CRES as a visiting fellow in 1976. Fenner and Costin associated with Geoff Mosely, a bushwalking admirer of Myles Dunphy and a wilderness enthusiast. They helped to set up the Australian Conservation Foundation, initially with funding entirely from the Federal Government. Fenner was vice president for a time and Mosely was Executive Director from 1973 to 1986. Fenner was a member of the Scientific Committee on Problems of the Environment that brought A. Malcolm Gill to prominence in fire ecology and created an ongoing dichotomy between theory and practice.

I have no doubt that Fenner was sincere in his concerns about human 'disturbance' of the environment. He stated in 2010: "Homo sapiens will become extinct, perhaps within 100 years … It's an irreversible situation. I think it's too late". Soon after, at the Australian Academy of Science, he opened the *Healthy Climate, Planet and People* symposium: "which is designed to bridge the gap between environmental science and policy".[8]

I believe that people without relevant scientific qualifications should not be appointed to official positions funded by taxpayers and able to influence environmental policy. It is bad enough that so many with recognised qualifications don't understand our environment and ecology.

Lydia and Stephen Pyne's view is very different to mine. They are experts in humanities, and believe that the industrial revolution has given man the overwhelming influence on biogeochemical cycles and climate. To them this decrees that environmentally sustainable management:

> is not a scientific matter. Engineers should know that technology can enable but not advise; scientists, that science can advise but not decide. Rather, the choice must reside in those other sloppy forms of inquiry and action that lie outside their purview.[9]

However I believe that scientific advice is not being heard because it has been drowned out by public officials, such as Fenner and Flannery, without scientific standing in the relevant fields. Another sometimes publicly funded protagonist of dangerous warming caused by man is Professor Will Steffen who, like Flannery, claims recent megafires in Australia as proof. I was surprised to find, on reading *The Last Lost World*, that Stephen Pyne, famous for his knowledge of global fire history, apparently shares Steffen's views on man-made climate change and "*The Great Acceleration*".[10]

Steffen and a couple of colleagues published an article in 2007 proposing:

> the Anthropocene, the current epoch in which humans and our societies have become a global geophysical force. The Anthropocene began around 1800 with the onset of industrialisation ... From a preindustrial value of 270–275 ppm, atmospheric carbon dioxide had risen to about 310 ppm by 1950. Since then ... the Great Acceleration ... from 310 to 380 ppm ... with about half of the total rise ... in just the last 30 years. The Great Acceleration is reaching criticality. Whatever unfolds, the next few decades will surely be a tipping point in the evolution of the Anthropocene.[11]

However the Pynes suggested that:

> Carbon dioxide veered away from the norm some 8,000 years ago and methane around 5,000. ... The most likely novel factor is a modern humanity that grasped more of the earth through agriculture. By slashing, draining, burning, and loosing domesticates

on reconstituted landscapes, people have released stored carbon and stimulated methane production. The liberated greenhouse gases have created a greenhouse effect that has steadily made the Earth into a self-steeping crock pot.[12]

Whereas, in *World Fire*, Stephen Pyne wrote:

The capture of fire by the genus Homo changed forever the history of the planet. ... no other human technology has influenced the planet for so long and so pervasively. A grand dialectic emerged between ... the earth's biota and ... humans, such that they co-evolved, welded by fire to a common destiny.[13]

Lack of any measurable global warming for a decade and a half after 1998 posed challenges for the "criticality" of *The Great Acceleration*, the surety of the "tipping point" and for the "self-steeping crock pot". But in any case, the pervasive influence of the firestick in evolution goes back to the hominids, and man's influence on carbon dioxide apparently goes back thousands of years.

The Pynes seem to me, torn between the evolutionary significance of man taking up the firestick, and the environmental significance of the Industrial Revolution. In my view, the introduction of the firestick to a new environment by Australia's original settlers was an earth-shattering development. That similar developments occurred around the globe at widely different times with different consequences; namely development of intensive agriculture and domestication of animals, or not; suggests to me that the Anthropocene cannot properly be a new epoch, nor can it have a hard boundary. Alternatively it must begin over a million years ago, when primitive man gained control of the balance of nature by taming fire.[14]

The pre-history and history of Australia suggest that the Industrial Revolution was not such a big deal as the kindling or dousing of the firestick. Bulldozers and chainsaws have had less impact on flora and fauna. The arrival of the firestick, and its dousing, both caused mass extinctions of plants and animals. Logging and mining haven't.

In any case, it is possible that ongoing natural climate change[15] will soon cool down the proponents of an Anthropocene. Forty years after *Time* and *Newsweek* first published articles raising alarms about global cooling, new scientific papers suggest that the world is on the brink of dramatic cooling due to reduced solar activity.[16] In 2014, Prime Minister Tony Abbott's chief business adviser warned that the country was ill-prepared to deal with the possible threat of global cooling. Greens' Deputy Leader, Adam Bandt, said that "Maurice Newman's latest statements are channelling the flat earth commentary of a lunatic fringe" and called on Abbott to dump Newman. The call was later echoed by former Climate Commissioner, turned Climate Councilor Will Steffen.

It matters not whether climate is actually warming or cooling, nor whether industrial emissions are having any influence. It is certain that man cannot control climate. However, man can eliminate megafires, enhance biodiversity and restore ecological resilience by using the firestick pragmatically and scientifically.

Endnotes

1 Flood 2006.

2 Keane and Crawley 2002.

3 Darwin 1860.

4 Buswell *et al.* 2011.

5 Moles 2012.

6 Australian Government 2002.

7 Ibid.

8 Jones 2010.

9 Lydia and Stephen Pyne 2012.

10 Ibid. p. 241-2.

11 Steffen *et al.* 2007.

12 op. cit. p. 230.

13 Pyne 1997.

14 Lydia and Stephen Pyne 2012.

15 Fegel 2009, Carter 2010.

16 Yamaguchi *et al.* 2010, Abdussamatov 2013.

14

IMAGINARY PROBLEMS AND REAL CRISES

Lock it up and let it burn don't let the cattle graze
Lock it up and let it burn when wildfires are ablaze
The conservation movement should hang their heads in shame
They lock it up and let it burn in conservation's name

Ernie Constance

This award winning song succinctly highlights that the "policies of those we pay to care, have led to devastation, extinctions and despair".[1] It was written after the 2003 alpine megafires by a grazier on the Monaro – a member of the "very inbred" community whose knowledge and experience went unheeded by Costin,[2] leading to the "flawed agenda" denounced in the song.

Ernie Constance eloquently portrayed the great respect of alpine pastoralists for their environment, and the social, economic and environmental damage consequent to the ascendancy of green bureaucrats. He mourned the death of cattle that could have been sustained by alpine pastures during the Millenium Drought. He mourned that megafires bared the mountains to erosion, reduced biodiversity and created eruptions of greenhouse gases. He highlighted the inevitable consequences of lightning strikes in unmanaged fuels.

In my view, this short ballad has scientific credibility that is lacking in the formal literature. Graziers in the High Country observed that grazing and burning protected soils and biodiversity and prevented megafires and massive erosion. They hypothesised that exclusion of grazing and burning would lead to megafires, increased erosion, and loss of biodiversity and water yield.[3] This hypothesis has been tested and confirmed by the outcomes of half a century of mismanagement by green bureaucracies.[4]

When Sam Clayton employed Costin he informed him that grazing and burning caused erosion and loss of biodiversity in the High Country. When Costin took up his position he blamed "burning and grazing" for killing snow gum woodlands and replacing them with dense heaths, or creating dense coppice scrubs.[5] It was actually megafires in 1926 and 1939 that caused this damage in the woodlands as well as severe erosion in the Alps. After grazing was excluded from 12,000 hectares of alpine country, the 1951/52 fires exacerbated the damage.[6] Costin saw the problems getting worse in 'protected' areas, but instead of re-examining his false assumptions, he invoked 'tipping points'.[7] Over half a century as an employee of SCS, CSIRO, as a visiting fellow at the Fenner School, and as a member of various advisory committees, he continued to campaign against grazing and burning.[8]

Surveyor Townsend reported that Aborigines burnt right across the mountains during their annual feasts on bogong moths, and that alpine ash trees were called black gums.[9] Costin thought that Aborigines burnt small areas, and that alpine ash ecosystems should have only one fire every century and a half. He said that "the big problem with all rangeland grazing – whether you start from an alpine area and go right through to a desert area or to a monsoon tropical area – is that the grazing is incredibly selective". Costin claimed there was "absolutely overwhelming evidence that rangeland grazing, particularly together with fire – which accentuates the pressure of high grazing selectivity – has no part in our higher mountain catchments".[10]

But his trials at Kosciuszko showed that a grazed and burnt plot had the same cover of live snow grass and herbs as its unburnt and grazed counterpart. There were 237 herbs per square metre in the burnt area compared to only 163 in the unburnt with 30% less cover. Unburnt areas had invading shrubs but the grazed and burnt area did not. Costin considered that "The effect of burning is not fully apparent from these quadrat results. The use of fire … is inadvisable, since it hastens the removal of the dead grass cover which offers some protection to the soil".[11]

I say that the trials showed exactly what the High Country graziers tried to explain to Costin – grazing without fire cannot control the accumulation of rank tussocks and shrubs that suppress the growth of stockfeed including all the more delicate herbs and grasses that contribute to biodiversity. Snowy Mountains Man, Tom Taylor, showed Costin and his colleagues that erosion occurred where rank tussock grasses were allowed to suppress low ground covers. Lessees repeatedly pointed out that accumulation of rank snow grass and/or shrubs fuelled wildfires that damaged catchments.[12] A recent, ongoing study at Barry Aitchison's Snowy Plains run has shown that burnt areas have greater species diversity and less shrubs than unburnt areas whilst grazing has had no impacts.[13]

Costin claimed that:

> hazard reduction burning ignores one of the most fundamental resource equations that exist with respect to our natural and near-natural lands: fuel = catchment protection = habitat. ... Of the threats we have been discussing, the hazard reduction burning is by far the most important. ... if we had the good professionals around in those resources that we are talking about – soil and water, wildlife et cetera – we would not see such nonsense.[14]

But graziers, foresters and some ecologists understand that high fuel loads suppress biodiversity and support high-intensity fires that lead to massive erosion. Price and Morgan found that "competitive dominance, whether it is by grasses, trees or shrubs, results in reductions in species richness".[15]

In 1836, Surveyor-General Mitchell had noted a diversity of groundcovers on the arid Darling plains compared to the Bogan where there were more trees and shrubs, but only a few very robust grasses.[16] Price and Morgan studied grassy woodlands that were 'protected' by declaring a national park in 1984. After fire was excluded for half a century and grazing was excluded for the last two decades, diversity was being choked out by tea tree scrub.[17] Studies of grazing[18] and burning[19] consistently show that biodiversity is increased when larger tussocks and/or shrubs are controlled by one or the other or both.

Costin told an interviewer in 2006:

Because of my continuing efforts in Kosciuszko, I became a member of the Kosciuszko National Park's advisory committee, and continued with that long after I left CSIRO. Also, in about the mid to late '60s, the New South Wales National Parks and Wildlife Service was set up under Tom Lewis, the Minister. ... And one of the first things he did was to ask Harry Frith and me, 'How can we best get a representative group of parks set up in New South Wales?' We told him he could do it quite quickly if he had two or three people on it [an advisory committee] whom we named ... This proceeded like a house on fire, and the meetings we had were very effective indeed – until every Tom, Dick and Harry in terms of interest groups [stakeholders?] wanted to get on the committee. In no time flat it got so bogged down that no decisions were possible because there was always something else that needed investigating, investigating, investigating.

One of the huge tragedies of that, from a conservation point of view, was the way the woodchip business got off the ground in New South Wales. Geoff Mosley was on that committee too, and Geoff and I had the job of attempting to define the coastal environments in New South Wales that merited coming into the National Parks and Wildlife system. We could do this quite readily from parks we knew, but to be on the safe side we then had to look at all the areas which were still Crown land, and because they were so important in all sorts of ways our recommendation simply was, 'Get the lot. Have all of the still available Crown lands.'

That was the time, however, when the meetings were starting to bog down, and that's when woodchips arrived. Not only state forests but other areas of uncommitted Crown land were brought into the woodchip system. And that's how New South Wales lost much of what could have been a magnificent belt of parkland, right down the coast of New South Wales.[20]

New South Wales now has that "belt of parkland, right down the coast" but it's not magnificent because it is not managed properly. Like Kosciuszko National Park it is periodically scourged by raging fires. Access for people has been lost or deliberately destroyed. Twenty percent of the area is being choked out by scrub whilst eucalypts are dying –

ravaged by pests, parasites and diseases.[21] Rare biota such as broad-headed snakes, smoky mice, and Imlay mallee are facing extinction. Common species such as grasstrees and smooth-barked apple are dying in waterfront parks[22] that should be jewels in Sydney's crown. Lives and houses are repeatedly lost in areas that should be healthy, open and safe environments.

Meanwhile, the green, left-wing journalists of the Australian Broadcasting Corporation are being funded by taxpayers to peddle junk science produced by anti-grazing, anti-burning, anti-logging advocates and Global Warming enthusiasts. (A survey of ABC journalists in 2013 revealed that the great majority of respondents vote for the Greens or the ALP.[23])

I complained to the ABC after they showed stark footage of clearfelling in an exotic pine plantation to illustrate a story about recommencement of selective harvesting in western Queensland native forests. This 'vision' was accompanied by the following 'audio bite':

> But Timber Queensland spokesman Rod McInnes says it is sensible to use a resource that has been locked up for years. 'The reality of life is that forest needs disturbance and forests which are selectively disturbed by logging, as we do, are going to be better forests in the fullness of time,' he said. 'Rather than those that are left alone to just grow rampantly without any management at all.'[24]

I had no difficulty visualising the undoubted sneer on the presenter's face. The ABC admitted that the report was factually incorrect and explained that they didn't have any real footage, nor the time to capture any, however they asserted that their report was balanced. The Australian Communications Management Authority supported this assertion as did the Commonwealth Ombudsman.

Allegedly balanced reporting such as this has empowered greens to dismantle native forestry in Australia, destroying local socioeconomies, or worse in the case of Tasmania. This is despite the fact that logging has not extinguished, or even endangered, a single species of plant or animal in this country. Lindenmayer claims that logging causes megafires and removes old trees, threatening Leadbeater's possum.[25] This is a complete distortion. It is megafires created by fire suppression[26] that have killed nearly all the old mountain ash trees and endangered the possum, just as they have the old snow gums in Kosciuszko where there is no logging. Lindenmayer and his colleagues in the Fenner School continue to advocate fire suppression, and if they have ongoing success, the next alpine megafire will probably extinguish Leadbeater's possum as well as the mountain pygmy possum and the corroboree frog.[27]

The ABC along with other green media and academia in Australia have been ably assisted by a gallery of celebrities in promoting their socioeconomically and environmentally destructive philosophies. Singers, actors and fashion designers such as Olivia Newton-John, Jimmy Barnes, Jack Thompson, Peter Garrett, Prue Acton and John Williamson easily spring to mind. As President of the Australian Conservation Foundation from 1989 to 1996, Garrett was well able to promote management by neglect. He was awarded the Order of Australia in 2003 "for service to the community as a prominent advocate for environmental conservation".[28] The ALP dumped a popular local member in 2004 to give him a safe seat in Federal Parliament. When they won government in 2007 he became Minister for the Environment. The Order of Australia citation mentioned services to both music and the environment. In my opinion Garrett was equally undeserving on both counts. I appreciate neither the green ideology in his lyrics nor his spastic on-stage antics.

British Royalty, in Prince Philip and Prince Charles, also push a green agenda. His Royal Highness Prince Philip wrote:

> I just wonder what it would be like to be … an animal … in danger of extinction. What would be its feelings toward the human species whose population explosion had denied it somewhere to exist … I am tempted to ask for reincarnation as a particularly deadly virus.[29]

Philip was instrumental in setting up the Australian Conservation Foundation with taxpayers funds, and was President from 1971 to 1976. In 1973 he allegedly engineered a 'coup' to radicalise the Board and establish wilderness enthusiast Geoff Mosley as Executive Director. During this time he was reportedly active against renewable energy schemes in Tasmania. He was alleged to have personally piloted the Royal jet over Lake Pedder and later engaged in a shouting match with 'Electric Eric' Reece the Premier of Tasmania.[30]

Prince Charles, on the other hand, is a global warming enthusiast who believes that "an unprecedented transformation of our communities, societies and lifestyles" will be needed to 'tackle' climate change. He apparently believes that this issue is more important than wars in the Middle East and Ukraine, and calls for renewable energy to be "vastly scaled up".[31] But there was some confusion in the media about whether Charles had expressed opposition to the Franklin Dam in 1983.[32]

When I was in secondary school, I achieved a Duke of Edinburgh Award, partly through developing skills and knowledge of the bush. I regret having mislaid the certificate and badge because I would enjoy burning the former and melting the latter now that I understand the green philosophies of the patron and his scion. I suppose that the Royal Prerogative of denying 'peasants' access to the game on their lavish estates is much the same as denying ordinary Australian people access to resources in the 'wilderness'. I think that British Royalty could do more good in the world by turning their estates, including thoroughbred studs and training tracks, to growing food for people. Prince Philip was awarded another Knighthood on Australia Day in 2015. Prime Minister Abbott referred to the good outcomes of the Duke of Edinburgh Award scheme in his announcement. I was as unimpressed as most other Australians including Mr Abbott's parliamentary colleagues who moved for a spill of the leadership.[33]

In 2014, on the opening day of the World Parks Congress in Sydney, Professor of Conservation Planning Bob Pressey and Senior Lecturer in Ecology Euan Ritchie posed the rhetorical question: "We have more parks than ever, so why is wildlife still vanishing?".[34] Their answer was that parks are leftover country that no-one wanted. They claimed that clearing, logging, grazing, fishing and mining are the 'culprits' behind loss of biodiversity. But virtually all of Australia's loss of biodiversity has been due to disruption of Aboriginal burning and lack of mild fire. Biodiversity has been lost from country that no-one wanted, like our sandstone national parks, as well as country that was taken out of sensible management like Kosciuszko.

When NPWS bought a farm to save the plains wanderer, they had to continue grazing to maintain its habitat. The Hastings River mouse was lost from open country when they turned it to scrub by excluding cattle and fire. No species has been lost through logging in Australia. It simply doesn't matter how many national parks are declared or where they are. Biodiversity continues to decline because lands are unmanaged or mismanaged with infrequent high-intensity wildfire.

John Williamson was admitted to the Order of Australia in 1992 with the following citation: "for service to Australian country music and in stimulating awareness of conservation issues".[35] Williamson sang the highly divisive song – *rip rip woodchip* – at the 1989 Rugby League Grand Final in Sydney:[36]

> What am I gonna do – what about the future?
> Gotta draw the line without delay
> Why shouldn't I get emotional – the bush is sacred
> Ancient life will fade away
>
> Over the hill they go, killing another mountain
> Gotta fill the quota – can't go slow
> Huge machinery wiping out the scenery
> One big swipe like a shearer's blow
>
> Rip rip woodchip – turn it into paper
> Throw it in the bin, no news today
> Nightmare, dreaming – can't you hear the screaming?
> Chainsaw, eyesore – more decay
>
> Remember the axemen knew their timber
> Cared about the way they brought it down
> Crosscut, blackbutt, tallowood and cedar
> Build another bungalow-pioneer town

I am the bush and I am koala
We are one – go hand in hand
I am the bush like Banjo and Henry
It's in my blood – gonna make a stand

Rip rip woodchip – turn it into paper
Throw it in the bin – don't understand
Nightmare, dreaming – can't you hear the screaming?
Stirs my blood – gonna make a stand

I've attempted to translate the song's message into English as follows:

- There is an environmental crisis requiring immediate action.
- I'm entitled to be emotional rather than scientific because I am of the wilderness religion.
- Logging will eliminate old trees.
- Logging quickly covers extensive areas and destroys forests and landforms.
- Logging produces only low value throw-away products, is noisy and unsightly, and promotes decay (presumably of trees).
- Oldtimers who 'mined' the bush without any thought for regeneration were morally superior to modern foresters because they were only cutting timber for houses.
- Logging endangers koalas and cultural heritage.
- My wants are more important than those of communities who rely on the timber industry for their livelihoods.

The message is wrong on all counts. Williamson denounced workers that cut and regenerate forests with their full complement of biodiversity, but he was proud that his forebears cleared and cropped The Mallee, dismantling entire ecosystems. Apart from wiping out some native plants and animals, they bared the ground and created dust storms that turned snows red in New Zealand, and fed phytoplankton in the Southern Ocean.[37] A verse from *Mallee Boy* reveals this arrogance and hypocrisy:

> Well I've ripped and dug out burrows on a sandy bulloak hill
> Eradicating rabbits doesn't take a lot of skill
> And a boy born in the Mallee doesn't find 'em hard to kill
> But they'll never be as rare as a Quandong tree
> My grandma made some jam for my brothers and me
> They're like the Mallee Fowl you hardly ever see
> But I don't mind at all if you call me a Mallee Boy.

The real environmental crises in Australia are not getting any attention because greens have irrational obsessions with carbon dioxide emissions, logging, prescribed burning and grazing. In 2013, during their first week of office, Australia's new coalition government under Tony Abbott attempted to eliminate a major distraction and waste of resources by sacking Flannery's Climate Commission. The ABC thwarted this example of democracy in action by using taxpayers' money to mount an advertising campaign for Flannery's rebirthed volunteer propaganda machine. They launched the campaign by inviting celebrity scientist and fellow devotee of Gaia, David Suzuki, to attest on national television that Tim and his Climate Commission were a vital asset to Australia. The campaign was successful, and the ABC continued to support the 'independent' climate council with free advertising:

> A climate science body abolished by the Federal Government has been relaunched as a community-funded organisation. The Climate Commission was set up to advise on the ... science and economics of carbon pricing, but was scrapped by the Government last week. The group's former chief commissioner, Tim Flannery, says thanks to enormous public support, it has been relaunched as the Climate Council. 'We need a clear, credible and authoritative and independent voice in this area and there has never really been a more critical time for that voice than now,' he said. ... 'We've been raising $1,000 an hour and that's through the night,' ... Under the previous government's model, the Climate Commission cost about

$5.4 million … over a four-year period. … Environment Minister
Greg Hunt told Lateline that the public support for the Climate
Council proves the Government should not have pay for body [sic].
'That's the great thing about democracy, it's a free country and it
proves our point that the commission didn't have to be a taxpayer
funded body,' he said. Mr Hunt said the Climate Commission
never did original research, instead it simply collated research from
other scientific agencies like the Bureau of Meteorology … But
Professor Flannery says that was the whole point of the Climate
Commission. 'We have simply one goal and one objective and we
always have, which is to take the science, the economics of climate
change and what's happening internationally in terms of action
and present it in a clear and understandable way and authoritative
way to the Australian public,' he said.[38]

I think it is a good thing that taxpayers are no longer funding a green,
left-wing propaganda machine to the tune of over a million dollars a
year. Unfortunately, the much bigger ABC is still going strong, and
costing taxpayers over a billion dollars every year. And the Australian
Bureau of Meteorology is still peddling propaganda. Not much has
changed. The Climate Commission advertised the *Angry Summer* of
2013, whereas the new Climate Council took the credit in 2014. It is
not science and the taxpayer is still covering the bill whilst the Climate
Council 'rattles the tip jar'.

The real environmental crisis in Australia is lack of mild fire and/or
grazing and/or thinning in native vegetation. Megafires, chronic eucalypt
decline, pestilence and loss of biodiversity are the consequences. Ernie
Constance deservedly received several prestigious awards such as Bush
Ballad of the Year for bringing this to public attention by singing *In
Conservation's Name*. Bill Gammage's service to Australian history has
been recognised with an AM and several literary prizes. Unfortunately
such honours have been devalued since Williamson and Garrett received
AMs and Flannery was proclaimed Australian of the Year.

Endnotes

1 Constance 2003.
2 Australian Academy of Science 2006.
3 Anon. 1969.
4 Jurskis *et al.* 2003, 2006; MCAV nd.
5 Costin 1954; Figs. 116, 121.
6 Jurskis *et al.* 2006.
7 Australian Academy of Science 2006.
8 Ibid.
9 Townsend 1846.
10 Australian Academy of Science 2006.
11 Costin 1954, pp. 377-84.
12 Anon. 1969, Merritt 2007.
13 Dr Lachlan Ingram pers. comm. 2015.
14 Australian Academy of Science 2006.
15 Price and Morgan 2010.
16 Mitchell 1839.
17 Price and Morgan 2008.
18 Price and Morgan 2007, 2008, 2010.
19 Penman *et al.* 2008, 2009; Jurskis 2011b; Jurskis and Underwood 2013.
20 Australian Academy of Science 2006.
21 Jurskis *et al.* 2003, Jurskis 2005, Jurskis 2011b, Jurskis and Walmsley 2011.
22 Cahill *et al.* 2008, Sydney Harbour Federation Trust 2004.
23 Hanusch 2013.
24 O'Brien and Withey 2013.
25 Lindenmayer 2012.
26 Attiwill *et al.* 2013.
27 Worboys 2003.
28 Australian Government nd.
29 HRH Prince Philip 1986.
30 Barwick 2013.
31 Gosden 2014.

32 Australian League of Rights 1983.

33 Safi 2015.

34 Pressey and Ritchie 2014.

35 op. cit.

36 Evans 2014.

37 *The Examiner* 1944, Bhattachan and D'Odorico 2014.

38 Borrello 2013.

14a

THE MAGIC PUDDING

'That's where the Magic comes in,' explained Bill.
The more you eats the more you gets'.

Norman Lindsay

Bill Gammage's complex and romantic view of firestick ecology seems to have burdened him with unnecessary pessimism. He concluded that:

> There is no return to 1788. Non-Aborigines are too many, too centralised, too stratified, too comfortable, too conservative, too successful, too ignorant. ... We take more and leave the future less. Too few accept that this behavior cannot survive the population time bomb. When the time comes to choose between parks and people, species and space, food and freedom, 1788's values will be obliterated.[1]

But Stephen Pyne has recognised that post-European Australia had already demonstrated that it was capable of restoring the environmental balance previously maintained by Aborigines. He described "firestick forestry" as "a singular achievement, unrivalled elsewhere in the developed world". The rise of the wilderness mentality destroyed that achievement and dragged us back to equality with the rest of the world, but we know that it can be done. Pyne wrote:

> This time – so far – no one has seized the firestick and redirected it to new purposes, what one might loosely categorise as firestick ecology. Instead the country seems locked into a polarised politics of identity for which conflagrations provide a dramatic backdrop.[2]

Hopefully, we are a little less ignorant since Gammage published his book. Maybe we can break out of the "polarised politics".

The Left could return to its working class roots, slough off the green slime that has infected them since the 1980s and once more embrace socialism. *The Biggest Estate on Earth* shows that we don't really need to "choose between parks and people". Indeed we need to put people back into parks so that we can **keep** "species and space, food and freedom". In not seeing how our *Magic Pudding* was made, wilderness enthusiasts also failed to see:

> The danger he is stood in
> Who steals a high explosive bomb,
> Mistaking it for Puddin'.[3]

Thus they have visited megafires, chronic decline of trees and biodiversity, and socioeconomic disruption upon us.

But the "puddin'-thieves" in conservation bureaucracies are slowly waking up to how it works. They are talking about doing more burning. They are grazing small areas of parks to maintain habitat for the plains wanderer. They are talking about ecological thinning. Of course, the only difference between ecological thinning and commercial thinning is that the former costs taxpayers a lot of money, wastes renewable energy, creates a fire hazard, obstructs access, makes an unsightly mess and degrades habitat, whereas the latter improves the environment, provides a return to taxpayers and reduces consumption of fossil fuels.

To conserve biodiversity and old trees, and to restore fire safety, conservation bureaucrats must understand that human economy made the pudding, and it needs to be eaten. As Lindsay wrote of the "Puddin'-Owners":

> They had a delightful meal, eating as much as possible, for whenever
> they stopped eating the Puddin' sang out –
> 'Eat away, chew away, munch and bolt and guzzle,
> Never leave the table till you're full up to the muzzle'

The *Puddin'* has sung out again to the tune of megafires, wall to wall scrub, dying trees and disappearing species. Our modern multiracial society can have a thriving ecology as well as a thriving economy, just as Aborigines did before we stole their *puddin'*. We only need to hear the *puddin'* singing out.

Endnotes

1 Gammage 2011.
2 Pyne 2006.
3 Lindsay 1918.

15

FIRESTICK ECOLOGY

*The Australia that is replacing hoofed beef and wool with
biodiversity needs not only to grab that firestick but to be educated
in its subtle deployment*

Stephen Pyne

All Australian ecosystems either depend on mild fire or live on sites that are physically protected from it. All, bar rainforests and scrubs, are fire-dependent. Even pockets of rainforest or scrub benefit from the protection afforded by mild fire in the landscape. Only truly wet eucalypt forests depend on intense fires. These will happen without human intervention, thanks to lightning storms. But without human intervention, megafires will reset all the small, disjunct areas of wet sclerophyll forest across whole regions to age zero. When this happens biodiversity is lost. This is why Leadbeater's possum is in danger of extinction.

Frequent mild fire is needed across the landscape to maintain natural nutrient cycling and the competitive balance amongst trees, arbivores, shrubs and herbs. Mild fire promotes the health of ancient trees and shrubs, including supposedly fire-sensitive obligate seeders, and favours herbs and grasses by killing nearly all woody seedlings, controlling shoot development from lignotubers, and burning litter and dry fallen timber. This maintains low levels of discontinuous fuels as well as sunny and airy conditions. When herbage or litter cures seasonally, fires can propagate under mild conditions.

Scrubs occur where the environment is too harsh to sustain trees and abundant grass. Lack of grass limits application of the firestick. Open scrubs such as mallee are accessible to the firestick after wet seasons produce a flush of herbage that turns into fuel as it cures, or after several seasons without fire allow fuel to accumulate. Thicker scrubs of large bushes can burn after litter accumulates over several seasons, but only when conditions are hot and dry enough to make the litter flammable, and there is enough wind to drive fire into the scrub. Closed scrubs only burn during drought and severe weather when fire can burn through canopies. Mild fires cannot penetrate swamps, rock outcrops, deep, dark gullies or closed scrubs because fuels are usually either too sparse or too damp. Fire-sensitive species grow naturally in these places where they have a competitive advantage over grasses, and often form a canopy that reinforces their inaccessibility to mild fire by shading.

Tall shrubs grow under tall eucalypt forests in sheltered position such as gullies and southerly aspects. Under warm, dry and windy conditions, fires burning from grassy forests penetrate these positions, igniting the litter and herbage and defoliating or killing some of the tall shrubs. The shrubs resprout from their branches or their stems or their bases, or from seeds released by woody fruits or buried in the soil.

Wet sclerophyll forests live in pockets that are nearly always too wet to carry fire even in hot, dry and windy weather, but are accessible to lightning fires or embers from nearby moist or dry ecosystems when extreme weather occurs during major droughts . Once every few centuries, a coincidence of drought, high temperatures, strong winds and lightning ignition or ember storms will allow fire to penetrate some of these areas. In a landscape maintained by the firestick, one or a few wet sclerophyll patches in a small part of any region will burn very fiercely. Most patches will remain unburnt because they are protected from ignition by a matrix of discontinuous fuels, by the short duration of extreme weather conditions and by the small number of extreme weather events each dry season.

Very infrequent, fierce fires in a patch of wet sclerophyll forest remove canopy, subcanopy and litter, and trigger massive eucalypt seedfall onto light and airy ashbeds. When it rains, eucalypt seeds germinate and quickly grow to saplings providing shade and shelter for a slower growing subcanopy of rainforest species. This is not succession. The physical dominance of eucalypts, the numerical dominance of rainforest species, and the boundaries with rainforest are maintained. The boundaries correspond to differences in soil and moisture conditions.

Rainforests occur where there is sufficient precipitation and/or shelter, and relatively deep, nutritious and well structured soils so that tall trees can grow and produce deep shade under a complete canopy. Dry fuels cannot accumulate on the ground under such a canopy, except in extreme droughts on a millennial scale. I didn't fully appreciate how much rainforests enhance their own habitat until I observed a number of celestial bodies in broad daylight from a small log dump within subtropical rainforest on the Koreelah Range above Wallaby Creek. Real rainforest is equally effective at excluding solar radiation from the floor as a deep mineshaft.

Frequent mild fire in the Australian landscape can't penetrate true rainforest. It maintains an open, seasonally flammable matrix within which physically protected or recently burnt, brown, black or green patches form a fine-grained mosaic supporting biodiversity. This fine, shifting mosaic breaks up a coarser landform matrix of grassland, scrub, savanna, forest and rainforest. Aborigines enhanced the fine and coarse-grained mosaics by burning according to season and site. Exposed areas were burnt before more sheltered areas because they could be. Brown, black or green patches weren't burnt because they couldn't be. By the time sheltered areas were hot and dry enough to burn they were surrounded by black or green areas so fires didn't escape.

This wasn't a complicated and highly skilled science developed over millennia, it was a basic science gained from participation. Aborigines burnt to make a living, they burnt to make war and they burnt to have fun. By the time children became adults they had a firm grasp of the firestick – they were masters of the physics and ecology of fire.

Stephen Pyne suggested that Firestick Ecology has to be more subtle than Firestick Forestry:

> An aboriginal model ... points to fire management based on control over ignition ... by suitable use of the firestick seeking to protect against the fires you don't want and promote the ones you

do. The strategy differs from hazard-reduction burning in both its philosophy and practice. The object is not, as in the former, simply better fire protection but a proper habitat.[1]

Perhaps Pyne was misled by academic fire ecologists. Fuel reduction improves habitat and increases biodiversity. Firestick Forestry as I knew it in 1980 was identical to the concept of Firestick Ecology that Pyne seeks to promote. Foresters Alan McArthur and Roy Free recognised in 1968 that grasslands and woodlands had to be burnt to remove rank grass and woody seedlings, promote herbs, and protect alpine bogs that provide habitat for sphagnum moss and corroboree frogs. The bogs have dried out and the frogs are threatened with extinction because megafires have been visited on them by green academics and bureaucrats who stole the firestick.

Burning by foresters and graziers maintained grassy edges against rainforest and dry rainforest at Wallaby Creek. Black-striped wallabies and two kinds of pademelons sheltered in the rainforests and fed close by, in grassy edges. The fauna was so rich because foresters maintained a rich mosaic of habitats. Since park managers doused the firestick in this World Heritage Area, feed for the rare macropods has been choked out by invading scrub, and old hollow trees that were homes for other rare animals are dying before their time.

The firestick can be an artist's brush when employed using the perspective of exposure and shade, fuel moisture differences, slope, wind speed and direction, humidity and temperature. Ignition pattern can be a work of art or, it can be a brutal assault, as in the Kapalga experiment and in typical, infrequent burning of national parks. Australian foresters developed Firestick Forestry in one generation and passed it on to baby boomers such as myself. We either let the firestick go out or had it stolen from us and doused by greens. Now they're telling us that things have changed too much and we couldn't rekindle it even if we wanted to, and in any case megafires are good for biodiversity!

Woody thickening and homogenisation of the landscape – loss of biodiversity – has indeed made it much harder to apply the firestick. National park managers routinely stuff up prescribed burns after long intervals (usually at least a decade) without fire. In the winter of 2000, four workers died in Ku-ring-gai Chase National Park when a prescribed burn went horribly wrong. In autumn 2005, 600 people had to be evacuated when a prescribed burn on Wilsons Promontory escaped and subsequently engulfed 6,000 hectares of National Park. In spring 2011, an escape into long-unburnt private property from prescribed burning in a long-unburnt National Park at Margaret River destroyed 39 properties causing tens of millions of dollars damage.

Incidents such as these are not a reason to stop burning. They are a reason to do more burning and to get it right. Long-unburnt fuels will inevitably burn in megafires. Frequent prescribed burning is essential for socioeconomic and environmental protection. The problem is that most fire ecologists and conservation bureaucrats haven't been brought up with the firestick and won't listen to those that were. But they don't even need to listen. If they opened their eyes and their minds they would see the problem and the solution.

Traditional Aborigines burnt all year round. When foresters took up the firestick they found that the more you do it, the easier it gets, and practice makes perfect if you learn from your mistakes. However they mostly didn't advance to the stage of burning in summer because they wished to protect timber quality in unnaturally dense and uniform young stands. Roger Underwood tells me that his generation of foresters used to burn moist karri (*E. diversicolor*) forests in midsummer because they couldn't be burnt in cooler and damper seasons.

As I edit this final chapter in February 2015, a huge conflagration that will become known as The Northcliffe Fire is burning out of control across more than 90,000 hectares of forest in the karri region of south-western Australia. Roger sent me an update by email, in which he 'wrote':

> This tragedy is the culmination of many years of mismanagement, mainly by the former DEC [Department of Environment and Conservation] who simply would not get on with … burning in the karri forest, intimidated by grape-growers and greenies and led astray by unaccountable academics. It was predictable … I

feel a great personal loss, as I ... initiated the move to reserve the beautiful virgin forests along the Boorara Brook and Canterbury River ... now all burnt to a cinder.

Managers of parks and reserves aren't constrained by the need to protect timber values. If they had the will and the skill they could reinstate perfectly safe and biodiverse natural ecosystems. Instead they constrain themselves with ridiculous theories about fire-sensitive plants and animals, and create unsafe, unhealthy, unnaturally homogenous landscape bombs. When they carry out prescribed burns at unnaturally long intervals, the heavy, three dimensionally continuous fuels together with their lack of practice and consequent ignorance of fire behaviour, ensure that ancient trees and shrubs are destroyed and thicker scrubs develop. With disuse or ignorant misuse of the firestick, rare species such as the broad-headed snake and Imlay mallee are heading for extinction, common species such as grasstrees and Tasmanian devils are threatened, and other species, both common and uncommon in nature, are pestilent.

David Bowman claimed that "Fire managers often see only 'fuel' and forget about the conservation of biodiversity so valued by ecologists. It remains a great challenge to balance these sharply contrasting but legitimate perspectives".[2] He doesn't understand that accumulation of fuel is actually homogenisation and loss of biodiversity. Ecologists who think otherwise don't understand ecology, let alone the physics of fire. There is only one legitimate perspective. The firestick created and maintained the biodiversity and the fire-safe environment that greeted European settlers. To conserve biodiversity and live safely, we need to apply it willingly, frequently and, with practice, skillfully.

Endnotes

1 Pyne 2006.
2 Bowman 2003.

ACKNOWLEDGEMENTS

Huiquan Bi, Pat Cooney and the late Jack Stewart gave me confidence to express my professional observations. Tom White encouraged me to publish them more widely in scientific journals. Peter Attiwill, Phil Cheney, John Turner and Roger Underwood inspired this book with their fundamentally important works as well as their generous criticisms and encouragement. David Packham helped it to happen, and Anthony Cappello's decision allowed it to happen. Barry Aitchison, Bob Bridges and Paul de Mar contributed substantially to the book, even before I started on it. My dearest wife Gwenny was very tolerant whilst I neglected my domestic responsibilities to write the book. My neighbour, Reidar Herfoss, supplied the quotation that headlines Chapter 6. He's a story on his own. I hope to personally thank all the other relatives, friends and colleagues that helped me to get it done.

BIBLIOGRAPHY

Abbott, I. 2003 "Aboriginal fire regimes in south-west Western Australia: evidence from historical documents". In *Fire in ecosystems of south-west Western Australia: impacts and management*. Eds. I. Abbott and N. Burrows. Backhuys Publishers, Leiden. pp. 119-46.

ABC 1998 *The Future Eaters* – ep3. http://www.abc.net.au/science/future/theses/theses3.htm

ABC News 2012 Koalas destroying Cape Otway manna gums. http://www.abc.net.au/news/2012-10-10/koalas-destroying-cape-otway-manna-gums/4305628

Abdussamatov, H.I. 2013 "Grand Minimum of the Total Solar Irradiance leads to the Little Ice Age". *Journal of Geology and Geosciences* 2, 113.

Adams, M., Attiwill, P. 2011 *Burning issues. Sustainability and management of Australia's southern forests*. CSIRO Publishing, Collingwood.

Andersen, A.N., Cook, G.D., Corbett, L.K., Douglas, M.M., Eager, R.W., Russell-Smith, J., Setterfield, S.A., Williams, R.J., Woinarski, J.C.Z. 2005 "Fire frequency and biodiversity conservation in Australian tropical savannas: implications from the Kapalga fire experiment". *Austral Ecology* 30, 155-67.

Anon. 1969 *A detailed examination and criticism by the Kosciusko National Park Snow Lessees and Occupiers Association of the "Report to the Honorable T.L. Lewis, M.L.A., Minister for Lands, on investigations into controlled grazing and longer term leases in the Kosciusko National Park" by Dr. Graham Edgar, O.B.E., D.V.Sc., A.R.C.V.S.*.

Anon. 1973 *Review of Major Fire Organisation – 1972/73 Fire Season.* Unpublished typescript. Forestry Commission of NSW.

Anon. 1989 *Forest Types in New South Wales*. Research Note No. 17. Forestry Commission of New South Wales, Sydney.

Attenborough, D. 1979 *Life on Earth*. Fontana/Collins.

Attiwill, P., Wilson, B. 2006 "Succession, disturbance and fire". In *Ecology: an Australian perpective*. Eds. P. Attiwill, B. Wilson. Oxford University Press. pp. 361-77.

Attiwill, P.M., Adams M.A. 2008 "Harnessing forest ecological sciences in the service of stewardship and sustainability: A perspective from 'down-under'". *Forest Ecology and Management* 256, 1636-45.

Attiwill, P.M., Packham, D., Barker, T., Hamilton, I. 2009 *The People's Review of Bushfires, 2002-2007, in Victoria. Final Report 2009.*

Attiwill, P.M., Ryan, M.F., Burrows, N., Cheney, N.P., McCaw, L., Neyland, M., Read, S. 2013 "Timber harvesting does not increase fire risk and severity in wet eucalypt forests of southern Australia". *Conservation Letters* 00, 1-14.

Auclair, A.N.D. 2005 "Patterns and general characteristics of severe forest dieback from 1950 to 1995 in the northeastern United States". *Canadian Journal of Forest Research* 35, 1342-55.

Auclair, A.N.D., Worrest, R.C., Lachance, D., Martin, H.C. 1992 "Climatic perturbation as a general mechanism of forest dieback". In *Forest Decline Concepts*. Eds. P.D. Manion, D. Lachance. APS Press, St. Paul, Minnesota. pp. 38-58.

AUSLIG 1990 *Atlas of Australian Resources, Third Series, Volume 6. Vegetation.* Australian Surveying and Land Information Group, Department of Administrative Services, Canberra.

Australian Academy of Science 2006 *Dr Alec Costin, alpine ecologist.* https://www.science.org.au/node/327991

Australian Bureau of Meteorology 2012 *Improving Australia's Climate Record.* http://www.bom.gov.au/water/news/article.php?id=63

Australian Bureau of Meteorology 2013 *Special Climate Statement 43 – extreme heat in January 2013.* http://www.bom.gov.au/climate/current/statements/scs43e.pdf

Australian Government Department of Industry 2002 *Prime Minister's Prize for Science Professor Frank Fenner, AC, CMG, MBE, FAA, FRS, FRACP, FRCP.* https://grants.innovation.gov.au/SciencePrize/Pages/Doc.aspx?name=previous_winners/PM2002Fenner.htm

Australian Government Department of the Environment nd a *Sarcophilus harrisii – Tasmanian Devil.* http://www.environment.gov.au/cgi-bin/sprat/public/publicspecies.pl?taxon_id=299

Australian Government Department of the Environment nd b *Wollemia nobilis – Wollemi Pine.* http://www.environment.gov.au/cgi-bin/sprat/public/publicspecies.pl?taxon_id=64545

Australian Government nd *It's an Honour.* http://www.itsanhonour.gov.au/honours/honour_roll/search.cfm?aus_award_id=1043291&search_type=advanced&showInd=true

Australian League of Rights 1983 "Behind the Franklin River Issue". In *On Target.* 21 January 1983. http://www.alor.org/Volume19/Vol19No1.htm

Avitabile, S.C., Callister, K.E., Kelly, L.T., Haslem, A., Fraser, L., Nimmo, D.G., Watson, S.J., Kenny, S.A., Taylor, R.S., Spence-Bailey, L.M., Bennett, A.F., Clarke, M.F. 2013 "Systematic fire mapping is critical for fire ecology, planning and management: A case study in the semi-arid Murray Mallee, south-eastern Australia". *Landscape and Urban Planning* 117, 81-91.

Banks, J.C.G. 1989 "A history of forest fire in the Australian Alps". In *The scientific significance of the Australian Alps. The proceedings of the first Fenner conference on the environment. September 1988.* Ed. R. Good. Australian Alps National Parks Liaison Committee and Australian Academy of Science, Canberra, Australia. pp. 265-80.

Barber, P., Paap, T., Burgess, T., Dunstan, W., Hardy, G. 2013 "A diverse range of Phytophthora species are associated with dying urban trees" *Urban Forestry & Urban Greening* 12, 569-75.

Barlow, B.A. 1981 "The Australian flora: its origin and evolution". In *Flora of Australia. Vol.1 Introduction.* Australian Government Publishing Service, Canberra. pp. 25-76.

Barwick, R. 2013 "Australian Conservation Foundation: The British Crown created green fascism". *EIR* March 1, 2013. pp. Economics 9-17.

Bathurst 1815 "Earl Bathurst to Governor Macquarie". In *Two Journals of Early Exploration in New South Wales.* A Project Gutenberg Australia eBook. http:// gutenberg.net.au/ebooks13/1300271h.html#ch-06

Baur, G.N. 1965, 1979 *Forest Types in New South Wales.* Research Note No. 17. Forestry Commission of N.S.W., Sydney.

Benson, J.S., Redpath, P.A. 1997 "The nature of pre-European native vegetation in south-eastern Australia: a critique of Ryan, D.G., Ryan, J.R., Starr, B.J. (1995) the Australian landscape – observations of explorers and early settlers". *Cunninghamia* 5, 285-328.

Bhattachan, A., D'Odorico, P. 2014 "Can land use intensification in the Mallee, Australia increase the supply of soluble iron to the Southern Ocean?" *Scientific Reports* 4, 6009.

Binns, D.L., Bridges, R.G. 2003 *Ecological Impacts and Sustainability of Timber Harvesting and Burning in Coastal Forests of the Eden Area. Establishment and Progress of the Eden Burning Study.* Technical Paper No. 67. Research and Development Division, State Forests of New South Wales, Sydney.

Birk, E.M., Bridges, R.G. 1989 "Recurrent fires and fuel accumulation in even-aged blackbutt (*Eucalyptus pilularis*) forests". *Forest Ecology and Management* 29, 59-79.

Blair, F. c1950 *Bogong High Plains and Victorian Alps.* Unpublished manuscript held by Mrs E. Brown, Mountbatten Avenue, Bright, Victoria. In Moriarty 1993.

Blaxland, G. 1823 *The journal of Gregory Blaxland, 1813 incorporating the journal of a tour of discovery across the Blue Mountains, New South Wales, in the year*

1813. A Project Gutenberg of Australia ebook. http://gutenberg.net.au/ebooks02/0200411.txt

Bobbink, R., Hicks, J., Galloway, J., Spranger, T., Alkemade, R., Ashmore, M., Bustamante, M., Cinderby, S., Davidson, E., Dentener, F., Emmett, B., Erisman, J.-W., Fenn, M., Gilliam, F., Nordin, A., Pardo, L., De Vries, W. 2010 "Global assessment of nitrogen deposition effects on terrestrial plant diversity: a synthesis". *Ecological Applications* 20, 30-59.

Boco Rock Wind Farm nd *Facts and Figures*. http://bocorockwindfarm.com.au/facts_figures

Boer, M.M., Sadler, R.J., Wittkuhn, R.S., McCaw, L., Grierson, P.F. 2009 "Long-term impacts of prescribed burning on regional extent and incidence of wildfires – Evidence from 50 years of active fire management in SW Australian forests". *Forest Ecology and Management* 259, 132-42.

Boland, D.J., Brooker, M.I.H., Chippendale, G.M., Hall, N., Hyland, B.P.M., Johnston, R.D., Kleinig, D.A., Turner, J.D. 1984. *Forest Trees of Australia*. Nelson CSIRO.

Bombala Times 2011 "Critically endangered tree receives a leg up on Mt Imlay". 27 Sept. 2011. http://www.bombalatimes.com.au/story/1103404/critically-endangered-tree-receives-a-leg-up-on-mt-imlay/

Bond, W.J. 2006 "The world is not as green as it could be and that requires explanation – a reply to White". *Journal of Vegetation Science* 17, 541-2.

Bond, W.J. 2008 "What limits trees in C₄ grasslands and savannas?" *Annual Review of Ecology, Evolution, and Systematics* 39, 641-59.

Bond, W.J., Keeley J.E. 2005 "Fire as a global 'herbivore': the ecology and evolution of flammable ecosystems". *TRENDS in Ecology and Evolution* 20, 387-94.

Borrello, E. 2013 "Tim Flannery relaunches scrapped Climate Commission as community-funded body". In *ABC News* 24 Sept. 2013. http://www.abc.net.au/news/2013-09-24/tim-flannery-to-relaunch-climate-commission/4976608

Botkin, D.B. 2012 *The Moon in the Nautilus Shell: Discordant Harmonies Revisited*. Oxford University Press.

Bowman, D.M.J.S. 1998. "The impact of Aboriginal Landscape burning on the Australian biota". *New Phytologist* 140, 385-410.

Bowman, D.M.J.S. 2000 *Australian Rainforests: islands of green in a sea of fire*. Cambridge University Press.

Bowman, D.M.J.S. 2003 "Bushfires: a Darwinian perspective". In *Australia*

Burning: Fire ecology, policy and management issues. Eds. G. Cary, D. Lindenmayer, S. Dovers. CSIRO Publishing.

Bradstock, R.A. 2008 "Effects of large fires on biodiversity in south-eastern Australia: disaster or template for diversity?" *International Journal of Wildland Fire* 17, 809-22.

Bradstock, R.A., Williams, R.J., Gill, A.M. 2012 "Future fire regimes of Australian ecosystems: new perspectives on enduring questions of management". In *Flammable Australia: Fire regimes, Biodiversity and Ecosystems in a changing world.* Eds. R.A. Bradstock, A.M. Gill, R.J. Williams. CSIRO Publishing. pp. 307-25.

Braithwaite, L.W., Turner, J., Kelly, J. 1984 "Studies on the arboreal marsupial fauna of eucalypt forests being harvested for woodpulp at Eden, NSW III. Relationship between faunal densities, eucalypt occurrence and foliage nutrients and soil parent materials". *Australian Wildlife Research* 11, 41-8.

Bridges, R.G. 2005. *Effects of logging and burning regimes on forest fuel in dry sclerophyll forests in south-eastern New South Wales. Initial results (1986–1993) from the Eden Burning Study Area.* Research Paper No. 40. Forest Resources Research. NSW Department of Primary Industries, Sydney.

Brogden, J. nd *My Story: John Brogden.* Lifeline. https://www.outoftheshadows.org.au/Home/My-Story/John-Brogden

Burrows, N. 2005 "Burning rocks". *Landscope* 20, 55-61.

Burrows, N.D., Burbridge, A.A., Fuller, P.J., Behn, G. 2006 "Evidence of altered fire regimes in the Western Desert region of Australia". *Conservation Science Western Australia* 5, 272-84.

Burrows, N.D., Ward, B., Robinson, A.D. 1995 "Jarrah forest fire history from stem analysis and anthropological evidence". *Australian Forestry* 58, 7-16.

Buswell, J.M., Moles, A.T., Hartley, S. 2011 "Rapid evolution is common in introduced plant species". *Journal of Ecology* 99, 214-24.

Cahill, D.M., Rookes, J.E., Wilson, B.A. Gibson, L., McDougall, K.L. 2008 "Phytophthora cinnamomi and Australia's biodiversity: impacts, predictions and progress towards control". *Australian Journal of Botany* 56, 279-310.

Calaby, J.H. 1966 *Mammals of the Upper Richmond and Clarence Rivers, New South Wales.* Division of Wildlife Research Technical Paper No. 10. Commonwealth Scientific and Industrial Research Organisation, Australia.

Calaby, J.H. 1983 "Kangaroos and their relatives". In *The Australian Museum Complete Book of Australian Mammals.* Ed. R. Strahan. Angus & Robertson Publishers. pp. 175-6.

Campbell, J.F. 1926 "Notes on 'Explorations under Governor Philip'". *The Royal Australian Historical Society. Journal and Proceedings* XII, 26-40.

Campbell, K.G., Hadlington, P. 1967 *The biology of the three species of phasmatids (Phasmatodea) which occur in plague numbers in forests of southeastern Australia.* Research Note No. 20, Forestry Commission of N.S.W., Sydney.

Carter, R.M. 2010 *Climate: The Counter Consensus.* Stacey International, London.

Catalyst 2012 *Richard Kingsford.* http://www.abc.net.au/catalyst/stories/3515118.htm

Chase, A. 1987 *Playing God in Yellowstone: The Destruction of America's First National Park.* Harvest Books, New York.

Chesterfield, E.A. 1986 "Changes in the vegetation of the river red gum forest at Barmah, Victoria". *Australian Forestry* 49, 4-15.

Citizens Electoral Council 2008 *The Words of Prince Philip, The Current Hitler.* http://cecaust.com.au/main.asp?sub=articles&id=phillipquotes.html

Clarke, M.F., Avitabile, S.C., Brown, L., Callister, K.E., Haslem, A., Holland, G.J., Kelly, L.T., Kenny, S.A., Nimmo, D.G., Spence-Bailey, L.M., Taylor, R.S., Watson, S.J., Bennett, A.F. 2010 "Ageing mallee eucalypt vegetation after fire: insights for successional trajectories in semi-arid mallee ecosystems". *Australian Journal of Botany* 58, 363-72.

Clarke, M.F., Schedvin, N. 1999 "Removal of bell miners *Manorina melanophrys* from *Eucalyptus radiata* forest and its effect on avian diversity, psyllids and tree health". *Biological Conservation* 88, 111-20.

Clarke, W.B. 1860 *Researches in the southern gold fields of New South Wales.* Reading and Wellbank, Sydney. https://books.google.com.au/books?id=cUk1AAAAMAAJ&printsec=frontcover#v=onepage&q&f=false

Clements, F.E. 1916 *Plant Succession; an analysis of the development of vegetation.* Publication No. 242. Carnegie Institute of Washington.

Close, D.C., Davidson, N.J., Johnson, D.W., Abrams, M.D., Hart, S.C., Lunt, I.D., Archibald, R.D., Horton, B., Adams, M.A. 2009 "Premature decline of *Eucalyptus* and altered ecosystem processes in the absence of fire in some Australian forests". *Botanical Review* 75, 191-202.

Close, D.C., Davidson, N.J., Swanborough, P.W. 2011a "Fire history and understorey vegetation: water and nutrient relations of *Eucalyptus gomphocephala* and *E. delegatensis* overstorey trees". *Forest Ecology and Management* 262, 208-14.

Close, D.C., Davidson, N.J., Swanborough, P.W., Corkrey, R. 2011b "Does low-intensity surface fire increase water- and nutrient-availability to overstorey *Eucalyptus gomphocephala?*" *Plant and Soil* 349, 203-14.

Close, D.C., Davidson, N.J., Watson, T. 2008 "Health of remnant woodlands in fragments under distinct grazing regimes". *Biological Conservation* 141, 2395-2402.

Coates, F., Lunt, I.D., Tremblay, R.L. 2006 "Effects of disturbance on population dynamics of the threatened orchid *Prasophyllum correctum* D.L. Jones and implications for grassland management in south-eastern Australia". *Biological Conservation* 129, 59-69.

Cocks, A.R. 1979 "Gum tree graveyard". *Sydney Morning Herald.* 20 September 1979. p. 6. https://news.google.com/newspapers?nid=1301&dat=1979092 0&id=_oRWAAAAIBAJ&sjid=eeYDAAAAIBAJ&pg=4206,7188145&hl= en

Cohen, T.J., Jansen, J.D., Gliganic, L.A., Larsen, J.R., Nanson, G.C., May, J-H., Jones, B.G., Price, D.M. 2015 "Hydrological transformation coincided with megafaunal extinction in central Australia". *GEOLOGY* 43, 195-8.

Cohn, J.S., Lunt, I.D., Ross, K.S., Bradstock, R.A. 2011 "How do slow-growing, fire-sensitive conifers survive in flammable eucalypt woodlands?" *Journal of Vegetation Science* 22, 425-35.

Collins, D. 1798. *An Account of the English Colony in New South Wales, Vol. 1 With Remarks On The Dispositions, Customs, Manners Etc. Of The Native Inhabitants Of That Country. To Which Are Added, Some Particulars Of New Zealand; Compiled, By Permission, From The Mss Of Lieutenant-Governor King.* Project Gutenburg Ebook. http://www.gutenberg.org/files/12565/12565-h/12565-h.htm#chapter8

Colloff, M.J. 2014 *Flooded Forest and Desert Creek: ecology and history of the river red gum.* CSIRO Publishing.

Colvin, M. 2014 "Bird numbers down in Australia's east as big dry takes its toll". In *PM.* http://www.abc.net.au/pm/content/2014/s4123210.htm

Commonwealth of Australia 1995 *National Forest Policy Statement: a new focus for Australia's forests,* second edition.

Concerned Scientists 2011. http://www.abc.net.au/environment/exclusives/highcountrygrazing.pdf

Connell, J.H., Slatyer, R.O. 1977 "Mechanisms of succession in natural communities and their role in community stability and organization". *American Naturalist* 111, 1119-44.

Conroy, R.J. 1996 "To burn or not to burn? A description of the history, nature and management of bushfires within Ku-ring-gai Chase National Park". *Proceedings of the Linnean Society of New South Wales* 116, 79-95.

Constance, E. 2003 "In Conservation's Name". In *A Dinkum Bushman's Hands.* LBS Music, Tamworth NSW.

Cook, E., Bird, T., Peterson, M., Barbetti, M., Buckley, B., D'Arrigo, R., Francey, R. 1992 "Climatic change over the last millennium in Tasmania reconstructed from tree-rings". *The Holocene* 2, 205-17.

Costin, A.B. 1954 *A study of the ecosystems of the Monaro Region of New South Wales.* A.H. Pettifer Government Printer, Sydney.

Cowles, H.C. 1899 "The ecological relations of the vegetation on the sand dunes of Lake Michigan". *Botanical Gazette* 27, 95-117 … 167-202.

Cowles, H.C. 1911 "The causes of vegetational cycles". *Annals of the Association of American Geographers* 1, 3-20.

Croft, M., Goldney, D., Cardale, S. 1997 "Forest and woodland cover in the central western region of New South Wales prior to European Settlement". In *Conservation Outside Nature Reserves.* Eds. P. Hale, D. Lamb. Centre for Conservation Biology, The University of Queensland. pp. 394-406.

Croke, J., Takken, S.M.I. 2005 *Soil Erosion Control and Mitigation on Forest Roads: Post-Bushfire Road Rehabilitation in the ACT Forest Estate.* Draft Final Report submitted to ACT Forests by the Forest Road Research Group, University of New South Wales.

Cronon, W. 1995 "The trouble with wilderness; or, getting back to the wrong nature". In *Uncommon ground: Toward Reinventing Nature.* Ed. W. Cronon. W. W. Norton & Company Inc.. pp. 69-90.

Cunningham, G.M., Mulham, W.E., Milthorpe, P.L., Leigh, J.H. 1981. *Plants of Western New South Wales.* NSW Government Printing Office.

Cunningham, T.M., Cremer, K.W. 1965 "Control of the understorey in wet eucalypt forests". *Australian Forestry* 29, 4-14.

Curby, P. 1997 *Forest History Project for State Forests of New South Wales Narrandera Study on Buckinbong, Gillenbah and Matong State Forests.* State Forests of New South Wales, Pennant Hills.

Curr, E.M. 1883 *Recollections of Squatting in Victoria, Then Called the Port Phillip District, from 1841 to 1851.* Facsimile Edition 1968. George Robertson/ Libraries Board of South Australia, Melbourne/Adelaide.

Dale, V.H., Cristafulli, C.M., Swanson, F.J. 2005 "25 Years of ecological change at Mount St. Helens". *Science* 308, 961-2.

Darwin, C. 1845 *The voyage of the Beagle.* http://www.gutenberg.org/cache/epub/944/pg944-images.html

Darwin, C. 1860 *The Origin of Species.* Signet Classics.

Dasmann, R.F. 1972 *Environmental Conservation*, Third Edition. John Wiley & Sons, Inc..

Dawkins, R. 2006 *The God Delusion.* Transworld Publishers.

Donovan, P. 1997 *A history of the Millewa Group of River Red Gum Forests.* State Forests of New South Wales.

Dureau de la Malle, A.J.C.A. 1825 "Mémoire sur l'alternance ou sur ce problème: la succession alternative dans la reproduction des espèces végétales vivant en société, est-elle une loi générale de la nature". *Annales des Sciences Naturelles* 15, 353-81.

Edgar, G. 1969 *Report to the Honorable T.L. Lewis, M.L.A., Minister for Lands, on investigations into controlled grazing and longer term leases in the Kosciusko National Park.*

Ellis, R.C. 1964 "Dieback of alpine ash in north eastern Tasmania". *Australian Forestry* 28, 75-90.

Ellis, R.C., Mount, A.B., Mattay, J.P. 1980 "Recovery of *Eucalyptus delegatensis* from high altitude dieback after felling and burning the understorey". *Australian Forestry* 43, 29-35.

Ellis, R.C., Pennington, P.I. 1992 "Factors affecting the growth of *Eucalyptus delegatensis* seedlings in inhibitory forest and grassland soils". *Plant and Soil* 145, 93-105.

Ellis, S., Kanowski, P., Whelan, R. 2004 *National Inquiry on Bushfire Mitigation and Management.* Council of Australian Governments, Canberra.

Elton, C.S. 1930 *Animal ecology and evolution.* Clarendon Press.

Enright, N.J., Lamont, B.B., Miller, B.P. 2005 "Anomalies in grasstree fire history reconstructions for south-western Australian vegetation". *Austral Ecology* 30, 668-73.

Environment Australia 2008 *Approved Conservation Advice for Eucalyptus imlayensis (Imlay Mallee).* http://www.environment.gov.au/biodiversity/threatened/species/pubs/12623-conservation-advice.pdf

Evans, G.W. 1814 "Assistant-Surveyor Evans' Journal 1813-1814". In *Two Journals of Early Exploration in New South Wales.* A Project Gutenberg Australia eBook. http://gutenberg.net.au/ebooks13/1300271h.html#ch-06

Evans, G.W. 1815 "The Journal of Assistant-Surveyor Evans". In *Two Journals of Early Exploration of New South Wales.* A Project Gutenberg Australia eBook. http://gutenberg.net.au/ebooks13/1300271h.html#ch-06

Evans, W. 2014 "The Best and Worst Grand Final Entertainment". *The New Daily.* http://thenewdaily.com.au/sport/2014/09/02/best-worst-grand-final-music-performances/

Fagg, P.C., Ward, B.K., Featherstone, G.R. 1986 "Eucalypt dieback associated with *Phytophthora cinnamomi* following logging, wildfire and favourable rainfall". *Australian Forestry* 49, 36-43.

Farr, J.D., Swain, D., Metcalf, F. 2004 "Spatial analysis of an outbreak of *Uraba lugens* (Lepidoptera: Noctuidae) in the south-west of Western Australia: does logging, vegetation type or fire influence outbreaks?" *Australian Forestry* 67, 101-13.

Fegel, G.F. 2009 "Earth on the brink of an Ice Age". *Pravda* 11 January 2009. http://english.pravda.ru/science/earth/11-01-2009/106922-earth_ice_age-1/

Fensham, R.J., Fairfax, R.J. 1996 "The disappearing grassy balds of the Bunya Mountains south-eastern Queensland". *Australian Journal of Botany* 44, 543-58.

Fensham, R.J., Fairfax, R.J. 2006 "Can burning restrict eucalypt invasion on grassy balds?" *Austral Ecology* 31, 317-25.

Ferguson, I., Cheney, P. 2011 "Wildfires, not logging, cause landscape traps". *Australian Forestry* 74, 362-5.

Fisher, J.T. 1983 "Water relations of mistletoes and their hosts". In *The Biology of Mistletoes*. Eds. M. Calder, P. Bernhardt. Academic Press, Sydney. pp. 161-84.

Flannery, T.F. 1994 *The Future Eaters*. Reed New Holland.

Flannery, T.F. 2005 *The Weather Makers: the history and future impact of climate change*. Text Publishing, Melbourne Australia.

Flannery, T.F. 2012 "After the future: Australia's new extinction crisis". *Quarterly Essay* 48, 1-80.

Flinders, M. 1814 *A Voyage to Terra Australis*. In *Matthew Flinders: Terra Australis*. Text Classics, 2012.

Flood, J. 2006 *The Original Australians: story of the Aboriginal people*. Allen & Unwin.

Florence, R. 2005 "Bell-miner-associated dieback: an ecological perspective". *Australian Forestry* 68, 263-6.

Florence, R.G. 1996 *Ecology and Silviculture of Eucalypt Forests*. CSIRO, Australia.

Floyd, A.G. 1989 *Rainforest Trees of Mainland South-eastern Australia*. Inkata Press.

Ford, H.A., Walters, J.R., Cooper, C.B., Debus, S.J., Doerr, V.A.J. 2009 "Extinction debt or habitat change? – ongoing loss of woodland birds in north-eastern New South Wales, Australia". *Biological Conservation* 142, 3182-90.

Forestry and Timber Bureau 1962 *Forest Trees of Australia*, Second Edition. Commonwealth of Australia.

Forestry Commission of N.S.W. 1985 *Management Plan for Murray Management Area*. Forestry Commission of N.S.W. Sydney.

Gammage, B. 2011 *The biggest estate on earth: how Aborigines made Australia*. Allen & Unwin, Sydney.

General Purpose Standing Committee No. 5 2012a *Report of Proceedings before General Purpose Standing Committee No. 5: Inquiry into Public Land Management. At Sydney on Tuesday 4 December 2012.* p. 39.

General Purpose Standing Committee No. 5 2012b *Report of Proceedings before General Purpose Standing Committee No. 5: Inquiry into Public Land Management. At Sydney on Friday 14 September 2012.* p. 23.

General Purpose Standing Committee No. 5 2015 *Wambelong Fire.* NSW Parliament. Legislative Council.

Geographical Names Board 1970 *Mount Twiss.* http://www.gnb.nsw.gov.au/place_naming/placename_search/extract?id=SXjLjzKmMn

Gibbons, P., Briggs, S.V., Ayers, D.A., Doyle, S., Seddon, J., McElhinny, C., Jones, M., Sims, R., Doody, J.S. 2008 "Rapidly quantifying reference conditions in modified landscapes". *Biological Conservation* 141, 2483-93.

Gilbert, J.M. 1959 "Forest succession in the Florentine Valley, Tasmania". *Papers and Proceedings of the Royal Society of Tasmania* 93, 129-51.

Gill, A.M. 1981 "Adaptive responses of Australian vascular plant species to fire". In *Fire and the Australian biota.* Eds. A.M. Gill, R.H. Groves, I.R. Noble. Australian Academy of Science, Canberra. pp. 243-72.

Gleason, H.A. 1926 "The Individualistic Concept of the Plant Association". *Bulletin of the Torrey Botanical Club* 53, 7-26.

Goodland, R.J. 1975 "The tropical origin of ecology: Eugen Warming's jubilee". *Oikos* 26, 240-5.

Gordon, G., Hrdina, F. 2005 "Koala and possum populations in Queensland during the harvest period 1906-1936". *Australian Zoologist* 33, 69-99.

Gosden, E. 2014 "Prince Charles: climate change is the greatest challenge facing humanity". *The Telegraph*, London, 22 September 2014. http://www.telegraph.co.uk/news/uknews/prince-charles/11110457/Prince-Charles-climate-change-is-the-greatest-challenge-facing-humanity.html

Gough, N. 1994 *Mud, Sweat and Snow: memories of Snowy workers 1949-1959,* Second Edition. Noel Gough, Tallangatta, Victoria.

Gould, J.S., McCaw, W.L., Cheney, N.P., Ellis, P.F., Knight, I.K., Sullivan, A.L. 2007 *Project Vesta – Fire in Dry Eucalypt Forest: Fuel structure, fuel dynamics and fire behavior.* Ensis-CSIRO, Canberra and Department of Environment and Conservation, Perth.

Hanusch, F. 2013 "Whose views skew the news? Media chiefs ready to vote out Labor, while reporters lean left". *The Conversation* 20 May 2013. http://theconversation.com/whose-views-skew-the-news-media-chiefs-ready-to-vote-out-labor-while-reporters-lean-left-13995

Happs, J. 2013 "Warmism's bellowing dinosaurs". *Quadrant Online* 2 May 2013. http://quadrant.org.au/opinion/doomed-planet/2013/05/warmism-s-bellowing-dinosaurs/

Haslem, A., Kelly, L.T., Nimmo, D.G., Watson, S.J., Kenny, S.A., Taylor, R.S., Avitabile, S.C., Callister, K.E., Spence-Bailey, L.M., Clarke, M.F., Bennett, A.F. 2011 "Habitat or fuel? Implications of long-term, post-fire dynamics for the development of key resources for fauna and fire". *Journal of Applied Ecology* 48, 247-56.

Hassell, C.W., Dodson, J.R. 2003 "The fire history of south-west Western Australia prior to European settlement in 1826-1829". In *Fire in Ecosystems of South-West Western Australia: Impacts and Management*. Eds. I. Abbott, N. Burrows. Backhuys, Leiden, The Netherlands. pp. 71-86.

Hateley, R. 2010 *The Victorian bush: its 'original and natural' condition*. Polybractea Press, South Melbourne.

Hawkesbury Institute for the Environment nd *EucFACE*. University of Western Sydney, Penrith. http://www.uws.edu.au/hie/facilities/EucFACE

Henderson, M.K., Keith, D.A. 2002 "Correlation of burning and grazing indicators with composition of woody understorey flora of dells in a temperate eucalypt forest". *Austral Ecology* 27, 121-31.

Hessburg, P.F., Agee, J.K., Franklin, J.F. 2005 "Dry forests and wildland fires of the inland northwest USA: Contrasting the landscape ecology of the pre-settlement and modern eras". *Forest Ecology and Management* 211, 117-39.

Hesterberg, G.A., Jurgensen, M.F. 1972 "The relation of forest fertilization to disease incidence". *Forestry Chronicle* 48, 92-96.

Hickey, J.E., Su,W., Rowe, P., Brown, M.J., Edwards, L. 1999 "Fire history of the tall wet eucalypt forests of the Warra ecological research site, Tasmania". *Australian Forestry* 62, 66-71.

Higgins, P.J., Peter, J.M., Steele, W.K. (Eds.) 2001 *Handbook of Australian, New Zealand and Antartic Birds. Volume 5: Tyrant-flycatchers to chats*. Oxford University Press, Melbourne.

Hornell, J. 1943 "Outrigger devices: distribution and origin". *Journal of the Polynesian Society* 52, 91-100.

Horton, B.M. 2012 "Mitigating the effects of forest eucalypt dieback associated with psyllids and bell miners in World Heritage Areas 2012 *Australasian Plant Conservation* 20, 11-3.

Horton, B.M., Glen, M., Davidson, N.J., Ratkowsky, D., Close D.C., Wardlaw, T.J., Mohammed, C. 2013 "Temperate eucalypt forest decline is linked to altered ectomycorrhizal communities mediated by soil chemistry". *Forest Ecology and Management* 302, 329-37.

Horton, D.R. 1982 "The burning question: Aborigines, fire and Australian ecosystems". *Mankind* 13, 237-51.

House of Representatives Select Committee 2003 *A Nation Charred: Inquiry into the Recent Australian Bushfires*. Parliament of the Commonwealth of Australia, Canberra.

Howitt, A.W. 1891 "The eucalypts of Gippsland". *Transactions of the Royal Society of Victoria* II, 81-120.

HRH Prince Philip 1986 "Foreword" In *If I were an Animal*. Ed. F. Cowles. William Morrow and Company, Inc., New York

Hudson, E.L.S. 1957 *Fire plan for the Hume-Snowy Fire District*. Hume-Snowy Catchment Areas Bush Fires Prevention Scheme.

Hunter, D.A., Speare, R., Marantelli, G., Mendez, D., Pietsch, R., Osborne, W. 2009 "Presence of the amphibian chytrid fungus *Batrachochytrium dendrobatidis* in threatened corroboree frog populations in the Australian Alps". *Diseases of Aquatic Organisms* 92, 209-16.

Hurditch, W.E., Hurditch, W.J. 1994 "The politics of bushfires: can we learn from past mistakes?" In *The burning continent. Forest ecosystems and fire management in Australia*. Ed. C. Ulyatt. Institute of Public Affairs, West Perth, Western Australia. pp. 37-51.

Jackson, W.D. 1968 "Fire, air, water and earth – an elemental ecology of Tasmania". *Proceedings of the Ecological Society of Australia* 3, 9-16.

Jackson, W.D. 1999 "The Tasmanian legacy of man and fire". *Papers and Proceedings of the Royal Society of Tasmania* 133, 1-14.

Jacobs, M.R. 1955 *Growth Habits of the Eucalypts*. A.J. Arthur, Commonwealth Government Printer, Canberra.

James, E.A., McDougall, K.L. 2007 "Extent of clonality, genetic diversity and decline in the endangered mallee *Eucalyptus imlayensis*". *Australian Journal of Botany* 55, 548-53.

Jansen, A., Robertson, A.I. 2001 "Relationships between livestock management and the ecological condition of riparian habitats along an Australian floodplain river". *Journal of Applied Ecology* 38, 63-75.

Jones, C. 2010 "Frank Fenner sees no hope for humans". *The Australian*, 16 June 2010. http://www.theaustralian.com.au/higher-education/frank-fenner-sees-no-hope-for-humans/story-e6frgcjx-1225880091722

Jones, M.E., Barmuta, L.A. 2000 "Niche differentiation among sympatric Australian dasyurid carnivores". *Journal of Mammalogy* 81, 434-47.

Jones, M.E., Shepherd, M., Henry, R.J., Delves, A. 2006 "Chloroplast DNA variation and population structure in the widespread forest tree, *Eucalyptus grandis*". *Conservation genetics* 7, 691-703.

Jones, R. 1969 "Fire-stick farming". *Australian Natural History*. September, 224-8.

Jurskis, V. 2002 "Restoring the prepastoral condition". *Austral Ecology* 27, 689-90.

Jurskis, V. 2004a "Foraging preference of the smoky mouse, *Pseudomys fumeus*, in south-eastern New South Wales: an examination of sampling strategies". *Australian Forestry* 67, 149-51.

Jurskis, V. 2004b. *Observations of eucalypt decline in temperate Australian forests and woodlands*. 2004 Gottstein Fellowship Report. Gottstein Trust, Melbourne. http://gottstein.nuttify.com/wp-content/blogs.dir/651/files/2012/09/vjurskis.pdf

Jurskis, V. 2005 "Eucalypt decline in Australia, and a general concept of tree decline and dieback". *Forest Ecology and Management* 215, 1-20.

Jurskis, V. 2008. "Drought as a factor in tree declines and diebacks". In *Droughts: Causes, Effects and Predictions*. Ed. J.M. Sanchez. Nova Science Publishers Inc., New York. pp. 331-41.

Jurskis, V. 2009 "River red gum and white cypress forests in south-western New South Wales, Australia: ecological history and implications for conservation of grassy woodlands". *Forest Ecology and Management* 258, 2593-601.

Jurskis, V. 2011a "Benchmarks of fallen timber and man's role in nature: some evidence from temperate eucalypt woodlands in southeastern Australia". *Forest Ecology and Management* 261, 2149-56.

Jurskis, V. 2011b "Human fire maintains a balance of nature". In *Proceedings of Bushfire CRC & AFAC 2011 Conference Science Day. 1 September 2011, Sydney Australia*. Ed. R.P. Thorton. Bushfire Cooperative Research Centre, Melbourne, Australia. pp. 129-38. http://www.bushfirecrc.com/resources/pages-129-138-human-fire-maintains-balance-nature

Jurskis, V. 2014 "Lost and found and lost and found: the discovery and rediscovery of Tench's Prospect Mount". *Journal of the Royal Australian Historical Society* 100, 144-57.

Jurskis, V., Bridges, B., de Mar, P. 2003 "Fire management in Australia: the lessons of 200 years". In *Joint Australia and New Zealand Institute of Forestry*

Conference Proceedings 27 April – 1 May 2003. Eds. E.G. Mason C.J. Perley. Ministry of Agriculture and Forestry, Wellington. pp. 353-68.

Jurskis, V., de Mar, P., Aitchison, B. 2006 "Fire management in the alpine region". In *Bushfire 2006. Life in a fire prone environment: translating science into practice*. Ed. C. Tran. Griffith University. CD.

Jurskis, V., Potter M. 1997 *Koala Surveys, Ecology and Conservation at Eden*. Research Paper No. 34. State Forests of New South Wales, Sydney.

Jurskis, V., Turner J. 2002 "Eucalypt dieback in eastern Australia: a simple model". *Australian Forestry* 65, 81-92.

Jurskis, V., Turner, J., Lambert, M., Bi, H. 2011 "Fire and N cycling: getting the perspective right". *Applied Vegetation Science* 14, 433-4.

Jurskis, V., Turner, R.J., Jurskis, D. 2005 "Mistletoes increasing in 'undisturbed' forest: a symptom of forest decline caused by unnatural exclusion of fire?" *Australian Forestry* 68, 221-6.

Jurskis, V., Underwood, R. 2013 "Human fires and wildfires on Sydney sandstones: History informs fire management". *Fire Ecology* 9, 8-24.

Jurskis, V., Walmsley T. 2011 *Eucalypt ecosystems predisposed to chronic decline: estimated distribution in coastal New South Wales*. http://www.bushfirecrc.com/publications/biblio/ag/J?sort=author&order=desc

Jurskis, V.P., Hudson, K.B., Shiels R.J. 1997 "Extension of the range of the smoky mouse *Pseudomys fumeus* (Rodentia: Muridae) into New South Wales with notes on habitat and detection methods". *Australian Forestry* 60, 99-101.

Kanowski, P. 2012 "The tragedy of Tasmania's forests: One act on from Flanagan". *Island Magazine*, 6 February 2012. http://tasmaniantimes.com/index.php/article/the-tragedy-of-tasmanias-forests-one-act-on-from-flanagan-

Keane, R.M., Crawley, M.J. 2002 "Exotic plant invasions and the enemy release hypothesis". *TRENDS in Ecology and Evolution* 17, 164-70.

Keith, D. 2004 *Ocean Shores to Desert Dunes. The Native Vegetation of New South Wales and The ACT*. Department of Environment and Conservation, Hurstville.

Keith, D.A. 2012 "Functional traits: their roles in understanding and predicting biotic responses to fire regimes from individuals to landscapes". In *Flammable Australia: Fire regimes, biodiversity and ecosystems in a changing world*. Eds. R.A. Bradstock, A.M. Gill, R.J. Williams. CSIRO Publishing. pp. 97-126.

Keith, D.A., Henderson, M.K. 2002 "Restoring the prepastoral condition: response to Jurskis". *Austral Ecology* 27, 691-3.

Keith, D.A., Williams, J.E., Woinarski, J.C.Z. 2002 "Fire management and biodiversity conservation: key approaches and principles". In *Flammable Australia: the Fire Regimes and Biodiversity of a Continent*. Eds R.A. Bradstock, J.E. Williams, A.M. Gill. Cambridge University Press. pp. 401-28.

Kershaw, A.P., Clark, J.S., Gill, A.M., D'Costa, D.M. 2002 "A history of fire in Australia". *In Flammable Australia: the Fire Regimes and Biodiversity of a Continent*. Eds. R.A. Bradstock, J.E. Williams, A.M. Gill. Cambridge University Press. pp. 3-25.

Lacey, C.J. 1973. *Silvicultural Characteristics of White Cypress*. Research Note No. 26. Forestry Commission of N.S.W..

Landsberg, J., Morse, J., Khanna, P. 1990. "Tree dieback and insect dynamics in remnants of native woodlands on farms". *Proceedings of the Ecological Society of Australia* 16, 149-65.

Langford, K.J., O'Shaughnessy, P.J. (Eds.) 1977 *First Progress Report, North Maroondah*. Melbourne and Metropolitan Board of Works, Catchment Hydrology Research Report, MMBW-W-0005.

Lawson, W. 1813 *Journal of an expedition across the Blue Mountains, 11 May – 6 June 1813*. State Library, New South Wales. http://acms.sl.nsw.gov.au/album/albumView.aspx?acmsID=447154&itemID=823646

Leaver, B., Good, R. 2004 "Fire values". In *An assessment of the values of Kosciuszko National Park*. Independent Scientific Committee. NSW National Parks and Wildlife Service. pp. 117-28.

Leichardt, F.W.L. 1843 Letter to Lt. Robert Lynd 19 February. In *The Letters of F.W. Leichardt Volume II*. Cambridge University Press. pp. 632-3.

Leopold, A. 1948 *A Sand County Almanac and Sketches here and there*. Oxford University Press, 1989.

Lewis, H.T. 1993 "In retrospect". In *Before the Wilderness: Environmental management by native Californians*. Eds. T.C. Blackburn, K. Anderson. Ballena Press. pp. 389-400.

Lindenmayer, D. 2012 "Sending Leadbeater's Possum down the road to extinction". *The Conversation*, 14 December 2012. http://theconversation.com/sending-leadbeaters-possum-down-the-road-to-extinction-11249

Lindenmayer, D.B., Hobbs, R.J., Likens, G.E., Krebs, C.J., Banks, S.C. 2011 "Newly discovered landscape traps produce regime shifts in wet forests". *Proceedings of the National Academy of Sciences USA* 108, 15887-91.

Lindenmayer, D.B., Laurance, W.F. 2012 "A history of hubris – cautionary lessons in ecologically sustainable forest management". *Biological Conservation* 151, 11-6.

Lindsay, N. 1918 *The Magic Pudding*. Angus & Robertson, 2010.

Lourenço, P., Medeiros, V., Gil, A., Silva, L. 2011 "Distribution, habitat and biomass of *Pittosporum undulatum*, the most important woody plant invader in the Azores Archipelago". *Forest Ecology and Management* 262, 178-87.

Low, T. 2003 *The New Nature: winners and losers in wild Australia*. Penguin Books.

Loyn, R.H., Runnals, R.G., Forward, G.Y., Tyers, J. 1983 "Territorial bell miners and other birds affecting populations of insect prey". *Science* 221, 1411-3.

Lunney D., Leary T. 1988 "The impact on native mammals of land-use changes and exotic species in the Bega district, New South Wales, since settlement". *Australian Journal of Ecology* 13, 67-92.

Lunney, D. 2001 "Causes of the extinction of native mammals of the Western Division of New South Wales: an ecological interpretation of the nineteenth century historical record". *Rangelands Journal* 23, 44-70.

Lunt, I. 2014 "The night the cold killed the mallee: extreme events & climate change". *Ecology for Australia*, 31 August 2014. http://ianluntecology.com/2014/08/31/frost/

Lunt, I.D. 1998 "*Allocasuarina* (*Casuarinaceae*) invasion of an unburnt coastal woodland at Ocean Grove, Victoria: structural changes 1971-1996". *Australian Journal of Botany* 46, 649-56.

Lunt, I.D., Eldridge, D.J., Morgan, J.W., Witt, G.B. 2007b "A framework to predict the effects of livestock grazing and grazing exclusion on conservation values in natural ecosystems in Australia". *Australian Journal of Botany* 55, 401-15.

Lunt, I.D., Jansen, A., Binns, D.L., Kenny, S.A. 2007a. "Long-term effects of exclusion of grazing stock on degraded herbaceous plant communities in a riparian *Eucalyptus camaldulensis* forest in south eastern Australia". *Austral Ecology* 32, 937-49.

MacPherson, M.A. 1886 "Some causes of the decay of the Australian forests". *Journal of Proceedings of the Royal Society of N.S.W.* XIX, 83-96.

Manion, P.D. 1991 *Tree Disease Concepts*, second ed. Prentice-Hall, New Jersey.

Manion, P.D. 2003 "Evolution of concepts in forest pathology". *Phytopathology* 93, 1052-55.

Manning, A.D., Lindenmayer, D.B., Barry, S.C., Nix, H.A. 2006 "Multiscale-scale site and landscape effects on the vulnerable superb parrot of south-eastern Australia during the breeding season". *Landscape Ecology* 21, 1119-33.

Manton, J.A. 1895 "Murray Red-gum Forest Reserves". In *Votes and Proceedings of the Legislative Assembly of New South Wales 1895*. Volume 4. pp. 1155-6.

Marohasy, J. 2012 *Plugging the Murray River's Mouth: The interrupted evolution of a barrier estuary*. Report No. 001/12. Australian Environment Foundation.

Marris, E. 2009 "Ecology: ragamuffin earth". *Nature* 460, 450-3.

Marshall, A.J. 1966 (Ed.) *The Great Extermination: A guide to Anglo-Australian cupidity wickedness & waste*. Heinemann.

Martin, R.W. 1985 "Overbrowsing and decline of a population of the koala, *Phascolarctos cinereus*, in Victoria. III. Population dynamics". *Australian Wildlife Research* 12, 377-85.

McAllister, P. 2010 *Pygmonia: In search of the secret land of the pygmies*. University of Queensland Press.

McArthur, A.J., Free, R.A. 1968 *Control burning policy in the Snowy Mountains of New South Wales*. Unpublished typescript. Forestry Commission of NSW, Sydney.

MCAV nd *Environmental Benefits of High Country Grazing*. http://www.mcav.com. au/pdf/submissions/environmental_benefits.pdf

McCallum, H. 2008 "Tasmanian devil facial tumour disease: lessons for conservation biology". *Trends in Ecology and Evolution* 23, 631-7.

McFadden, M. 2013 "Australian endangered species: Southern Corroboree Frog". *The Conversation*, 25 July 2013. http://theconversation.com/au

McIntosh, P.D., Laffan, M.D., Hewitt, A.E. 2005 "The role of fire and nutrient loss in the genesis of the forest soils of Tasmania and southern New Zealand". *Forest Ecology and Management* 220, 185-215.

Meiklejohn, K.A., Dowton, M., Pape, T., Wallmam, J.F. 2013 "A key to the Australian Sarcophagidae (Diptera) with special emphasis on *Sarcophaga* (*sensu lato*)". *Zootaxa* 3680 (1). http://biotaxa.org/Zootaxa/article/view/zootaxa.3680.1.11

Merritt, J. 2007 *Losing Ground: Grazing in the Snowy Mountains 1944-1969*. Turalla Press.

Middleton, A. 2010 "River red gums gain new protection". *Australian Geographic*, 20 May 2010.

Miller, A.C., Watling, J.R., Overton, I.C., Sinclair, R. 2003 "Does water status of *Eucalyptus largiflorens* (Myrtaceae) affect infection by the mistletoe *Amyema miquelii* (Loranthaceae)?" *Functional Plant Biology* 30, 1239-47.

Miller, G.H., Fogel, M.L., Magee, J.W., Gagan, M.K., Clarke, S.J., Johnson, B.J. 2005 "Ecosystem collapse in Pleistocene Australia and a human role in megafaunal extinction". *Science* 309, 287-90.

Mills, K. 1988 "The clearing of the Illawarra rainforests: problems in reconstructing pre-European vegetation patterns". *Australian Geographer* 19, 230-40.

Mitchell, T.L. 1839 *Three Expeditions into the Interior of Eastern Australia; With Descriptions of the Recently Explored Region of Australia Felix and of the Present Colony of New South Wales,* Second ed., Rediscovery Books, Uckfield, Facsimile Edition 2006.

Mitchell, T.L. 1848 *Journal of an Expedition into the Interior of Tropical Australia in Search of a Route from Sydney to the Gulf of Carpentaria.* Longman, Brown, Green and Longmans, London. Facsimile Edition 2007. Archive CD Books Australia.

Moles, A.T., Flores-Moreno, H., Bonser, S.P., Warton, D.I., Helm, A., Warman, L., Eldridge, D.J., Jurado, E., Hemmings, F.A., Reich, P.B., Cavander-Bares, J., Seabloom, E.W., Mayfield, M.M., Shiel, D., Djietror, J.C., Peri, P.L., Enrico, L., Cabido, M.R., Setterfield, S.A., Lehmann, C.E.R., Thomson, F.J. 2012 "Invasions: the trail behind, the path ahead, and a test of a disturbing idea". *Journal of Ecology* 100, 116-27.

Montreal Process Implementation Group for Australia and National Forest Inventory Steering Committee 2013. *Australia's State of the Forests Report 2013.* ABARES, Canberra, December.

Mooney, S.D., Harrison, S.P., Bartlein, P.J., Daniau, A.-L., Stevenson, J., Brownlie, K.C., Buckman, S., Cupper, M., Luly, J., Black, M., Colhoun, E., D'Costa, D., Dodson, J., Haberle, S., Hope, G.S., Kershaw, P., Kenyon, C., McKenzie, M., Williams, N. 2011 "Late quaternary fire regimes of Australia". *Quaternary Science Reviews* 30, 28-46.

Mooney, S.D., Harrison, S.P., Bartlein, P.J., Stevenson, J. 2012 "The prehistory of fire in Australia". In *Flammable Australia: fire regimes, biodiversity and ecosystems in a changing world.* Eds. R.A. Bradstock, A.M. Gill, R.J. Williams. CSIRO Publishing. pp. 3-26.

Moore, K.M. 1961 "Observations on some Australian forest insects 8. The biology and occurrence of *Glycaspis baileyi* Moore in New South Wales". *Proceedings of the Linnean Society of New South Wales* 86, 185-200.

Moriarty, O.M. 1993 *Conservation or destruction of natural resources of alpine regions of south eastern Australia.* Oliver Moriarty, Aldgate S.A..

Mount, A.B. 1964 "The interdependence of the eucalypts and forest fires in southern Australia". *Australian Forestry* 28, 166-72.

Mueller, F. 1855 "II. – Vegetation". In *Researches in the Southern Gold Fields of New South Wales.* Ed. W.B. Clarke. Reading and Wellbank, Sydney, 1860. pp. 231-4

Murphy, L. 1983 In *Commonwealth v Tasmania ("Tasmanian Dam case") [1983] HCA 21; (1983) 158 CLR 1 (1 July 1983).*

Murray-Darling Basin Authority nd *Basin Plan.* http://www.mdba.gov.au/what-we-do/basin-plan

National Archives of Australia 2015 *The Age tapes and the investigation of Justice Lionel Murphy.* http://www.naa.gov.au/collection/explore/cabinet/by-year/1984-85/the-age.aspx

Natural Resource Management Ministerial Council 2009 *National Koala Conservation and Management Strategy 2009-2014.* Department of the Environment, Water, Heritage and the Arts, Canberra.

Natural Resources Commission 2009 *Final Assessment Report. Riverina Bioregion Regional Forest Assessment: River red gums and woodland forests.* Natural Resources Commission, Sydney, December 2009.

Neidjie, B. 1989 *Story about feeling.* Magabala Books, Broome, Western Australia.

Newsome, A.E. 1983 "Dingo" In *Complete Book of Australian Mammals.* Ed. R. Strahan. Angus & Robertson Publishers. p. 483.

Noble, J.C. 1997 *The delicate and noxious scrub: CSIRO studies on native tree and shrub proliferation in the semi-arid woodlands of eastern Australia.* CSIRO Division of Wildlife and Ecology, Lyneham, Australian Capital Territory, Australia.

Norton, A. 1887 "On the decadence of Australian forests". Proceedings of the Royal Society of Queensland 1886 III, 15-22.

NSW Government 1814 "Government Order, Sydney, Feb. 12, 1814". In *Project Gutenberg of Australia eBook.* http://gutenberg.net.au/ebooks02/0200411.txt

NSW Office of Environment & Heritage 2011 *Case study 4: Threatened species management in Oolambeyan National Park.* http://www.environment.nsw.gov.au/sop04/sop04cs4.htm

NSW Office of Water 2010 *Returning environmental flows to the Snowy River: An overview of water recovery, management and delivery of increased flows.* http://www.water.nsw.gov.au/__data/assets/pdf_file/0008/549170/snowy_initiative_returning_environmental_flows_snowy_river.pdf

NSWEPA nd *Terms of Licence under the Threatened Species Conservation Act 1995: Upper North East Region.* http://www.epa.nsw.gov.au/resources/forestagreements/UNETSLam7.pdf

O'Brien, C., Withey, A. 2013 "Qld's state forests reopen to logging". In *ABC News,* 27 February 2013. http://www.abc.net.au/news/2013-02-27/qlds-state-forests-reopen-to-logging/4542362

Old, K.M. 2000. "Eucalypt diseases of complex etiology". In *Diseases and Pathogens of Eucalypts*. Eds. P.J. Keane, G.A. Kile, F.D. Podger, B.N. Brown. CSIRO, Melbourne. pp. 411-25.

Ostry, M.E., Venette, R.C., Juzwik, J. 2010 "Decline as a Disease Category: Is It Helpful?" *Phytopathology* 101, 404-9.

Oxley, J.J.W.M. 1820 *Journals of Two Expeditions into the Interior of New South Wales Undertaken by Order of the British Government in the years 1817-18*. John Murray/ University of Sydney Library, London/Sydney, Etext 2002.

Palzer, C. 1981. "Eucalypt diebacks in forests". In *Eucalypt Dieback in Forests and Woodlands*. Eds. K.M. Old, G.A. Kile, C.P. Ohmart. CSIRO, Australia. pp. 9-12.

Parks & Wildlife Service Tasmania 2014 *Tasmanian Devil, Sarcophilus harrisii*. http://www.parks.tas.gov.au/?base=387

Parliament of New South Wales 2010 *National Park Estate (Riverina Red Gum Reservations) Bill 2010 (No. 2)*. https://www.parliament.nsw.gov.au/prod/ PARLMENT/hansArt.nsf/0/AD1C2BF5B2E3F264CA257734000E4ABA

Parliament of Victoria 2015 *Parliamentary Debates (Hansard). Legislative Council. Fifty-eighth Parliament, First Session, Tuesday 5 May 2015*. (Extract from book 6).

Parris, H.S. 1948 "Koalas on the lower Goulburn". *Victorian Naturalist* 64, 192-3.

Parsons, R.F., Uren, N.C. 2011 "Is Mundulla Yellows really a threat to undisturbed native vegetation? A comment". *Ecological Management & Restoration* 12, 72-3.

Parsons, R.F., Uren, N.C. 2011 "The relationship between lime chlorosis, trace elements and Mundulla Yellows". *Australasian Plant Pathology* 36, 415-18.

Pavlic, D., Barber, P., Slippers, B., Hardy, G., Wingfield, M., Burgess, T. 2008 "Seven new species of the *Botryosphaeriaceae* from baobab and other native trees in Western Australia". *Mycologia* 100, 851-66.

Pearlman, J., Clennell, A. 2005 "Brogden's parting swipe at Lib enemy". *The Sydney Morning Herald*, 30 August 2005.

Penman, T.D., Binns, D.L., Brassil, T.E., Shiels, R.J., Allen, R.M. 2009 "Long-term changes in understorey vegetation in the absence of wildfire in south-east dry sclerophyll forests". *Australian Journal of Botany* 57, 533-40.

Penman, T.D., Binns, D.L., Shiels, R.J., Allen R.M., Kavanagh, R.P. 2008 "Changes in understorey plant species richness following logging and prescribed burning in shrubby dry sclerophyll forests of south-eastern Australia". *Austral Ecology* 33, 197-210.

Penman, T.D., Kavanagh, R.P., Binns, D.L., Melick, D.R. 2007 "Patchiness of prescribed burns in dry sclerophyll eucalypt forests in south-eastern Australia". *Forest Ecology and Management* 252, 24-32.

Petraitis, P. 2013 *Multiple Stable States in Natural Ecosystems.* Oxford University Press.

Pickett, S.T.A., White, P.S. (Eds.) 1985. *The Ecology of Natural Disturbance and Patch Dynamics.* Academic Press.

Pimm, S.L. 1991 *The Balance of Nature: Ecological issues in the conservation of species and communities.* The University of Chicago Press.

Podger, F.D. 1981 "Some difficulties in the diagnosis of drought as the cause of dieback". In *Eucalypt Dieback in Forests and Woodlands.* Eds. K.M. Old, G.A. Kile, C.P. Ohmart. CSIRO, Australia. pp. 167-73.

Poiani, A. 1993 "Bell miners: what kind of farmers are they?" *Emu* 93, 188-94.

Pratt, B.H., Heather, W.A. 1973 "The origin and distribution of *Phytophthora cinnamomi* Rands in Australian native plant communities and the significance of its association with particular plant species". *Australian Journal of Biological Science* 26, 559-73.

Pressey, B., Ritchie, E. 2014 "We have more parks than ever, so why is wildlife still vanishing?" *The Conversation*, 12 November 2014. http://theconversation.com/we-have-more-parks-than-ever-so-why-is-wildlife-still-vanishing-34047

Price, J. 1798 *Journey into the interior of the country New South Wales, 24 January-2 February 1798 and 2nd Journey, 9 March-2 April 1798, by John Price.* Unpublished Manuscript. State Library of NSW. http://www.sl.nsw.gov.au/discover_collections/history_nation/exploration/early/interior/price.html

Price, J.N., Morgan, J.W. 2008 "Woody plant encroachment reduces species richness of herb-rich woodlands in southern Australia". *Austral Ecology* 33, 278-89.

Price, J.N., Morgan, J.W. 2009 "Multi-decadal increases in shrub abundance in non-riverine red gum (*Eucalyptus camaldulensis*) woodlands occur during a period of complex land-use history". *Australian Journal of Botany* 57, 163-70.

Price, J.W., Morgan, J.W. 2010 "Small-scale patterns of species richness and floristic composition in relation to microsite variation in herb-rich woodlands". *Australian Journal of Botany* 58, 271-79.

Prideaux, G.J., Gully, G.A., Couzens, A.M.C., Ayliffe, L.K., Jankowski, N.R., Jacobs, Z., Roberts R.G., Hellstrom, J.C., Gagan, M.K., Hatcher, L.M. 2010 "Timing and dynamics of Late Pleistocene mammal extinctions in southwestern Australia". *PNAS* 107, 22157-62.

Pringle, R.M., Syfert, M., Webb, J.K., Shine, R. 2009 "Quantifying historical changes in habitat availability for endangered species: use of pixel- and object-based remote sensing". *Journal of Applied Ecology* 46, 544-53.

Prober, S.M., Lunt, I.D., Thiele, K.R. 2002a "Determining reference conditions for management and restoration of temperate grassy woodlands: relationships among trees, topsoils and understorey flora in little-grazed remnants. *Australian Journal of Botany* 50, 687-97.

Prober, S.M., Thiele, K.R., Lunt, I.D. 2002b "Identifying ecological barriers to restoration in temperate grassy woodlands: soil changes associated with different degradation states". *Australian Journal of Botany* 50, 699-712.

Pugh, C., Blashki, P. 1989 *A Kingdom Lost: A story of the devastation of our wilderness.* William Heinemann, Australia.

Pugh, D. 2014 *For Whom the Bell Tolls.* North East Forest Alliance.

Pyne, L.V., Pyne, S.J. 2012 *The Last Lost World.* Viking.

Pyne, S. 2006 *The Still-Burning Bush.* Scribe Short Books, Melbourne.

Pyne, S.J. 1991 *Burning bush. A fire history of Australia.* University of Washington Press, Seattle.

Pyne, S.J. 1997 *World Fire: The culture of fire on Earth.* University of Washington Press.

Reid, N., Yan, Z. 2000 "Mistletoes and other phanerogams parasitic on eucalypts". In *Diseases and Pathogens of Eucalypts.* Eds. P.J. Keane, G.A. Kile, F.D. Podger, B.N. Brown. CSIRO, Melbourne. pp. 353-84.

Robertson, A.I., Rowling, R.W. 2000 "Effects of livestock on riparian zone vegetation in an Australian dryland river". *Regulated Rivers: Research and Management* 16, 527-41.

Robinson, R.T. 1997 *Dynamics of coarse woody debris in floodplain forests: impacts of forest management and flood frequency.* Honours Dissertation. Charles Sturt University, Wagga Wagga.

Rolls, E. 1981 *A Million Wild Acres.* Thirtieth Anniversary Edition, 2011. Hale & Iremonger, McMahon's Point.

Rose, S. 1997 "Influence of suburban edges on invasion of *Pittosporum undulatum* into the bushland of northern Sydney, Australia". *Australian Journal of Ecology* 22, 89-99.

Routley, R., Routley, V. 1973 *The fight for the forests : the takeover of Australian forests for pines, wood chips and intensive forestry.* Australian National University, Research School of Social Sciences, Canberra.

Ruhlen, M. 1987 *A Guide to the World's Languages: Volume 1. Classification*. Stanford University Press.

Rule, S., Brook, B.W., Haberle, S.G., Turney, C.S.M., Kershaw, A.P., Johnson, C.N. 2012 "The Aftermath of Megafaunal Extinction: Ecosystem transformation in Pleistocene Australia". *Science* 335, 1483-6.

Russell-Smith, J., Cook, G.D., Cooke, P.M., Edwards, A.C., Lendrum, M., Meyer, C.P., Whitehead, P.J. 2013 "Managing fire regimes in north Australian savannas: applying Aboriginal approaches to contemporary global problems". *Frontiers in Ecology and Environment* 11, e55-63.

Ryan, B. 2011 "Newman pledges abolishing Wild Rivers laws". In *ABC News*, 4 November 2011. http://www.abc.net.au/news/2011-11-04/newman-pledges-abolishing-wild-rivers-laws/3629476

Safi, M. 2015 "How giving Prince Philip a knighthood left Australia's PM fighting for survival". *The Guardian*, 3 February 2015. http://www.theguardian.com/australia-news/2015/feb/03/how-giving-prince-philip-a-knighthood-left-australias-pm-fighting-for-survival

Scott, P., Burgess, T., Barber, P., Shearer, B., Stukely, M., Hardy, G., Jung, T. 2009 "*Phytophthora multivora sp. nov.*, a new species recovered from declining *Eucalyptus*, *Banksia*, *Agonis* and other plant species in Western Australia". *Persoonia: a mycological journal* 22, 1-13.

Simberloff, D. 2014 "The 'Balance of Nature' – Evolution of a panchreston". *PLOS Biology* 12, e1001963.

Sinclair, S.J. 2006 "The influence of dwarf cherry (*Exocarpos strictus*) on the health of river red gum (*Eucalyptus camaldulensis*)". *Australian Forestry* 69, 137-41.

Singh, G., Kershaw, A.P., Clark, R. 1981 "Quaternary vegetation and fire history in Australia". In *Fire and the Australian biota*. Eds. A.M. Gill, R.H. Groves, I.R. Noble. Australian Academy of Science, Canberra, Australia.

Snowyhydro nd *Snowy Mountains Scheme*. http://www.snowyhydro.com.au/energy/hydro/snowy-mountains-scheme/

Somerville, S., Somerville, W., Coyle, R. 2011 "Regenerating native forest using splattergun techniques to remove lantana". *Ecological Management and Restoration* 12, 164-74.

Speer, M.S., Leslie, L.M., Colquhoun, J.R., Mitchell, E. 1996 "The Sydney, Australia wildfires of January 1994 – meteorological conditions and high resolution modeling experiments". *International Journal of Wildland Fire* 6, 145-54.

Stanton, P., Parsons, M., Stanton, D., Stott, M. 2014a "Fire exclusion and the changing landscape of Queensland's Wet Tropics Bioregion 2. The dynamics of transition forests and implications for management". *Australian Forestry* 77, 58-68.

Stanton, P., Stanton, D., Stott, M., Parsons, M. 2014b "Fire exclusion and the changing landscape of Queensland's Wet Tropics Bioregion 1. The extent and pattern of transition". *Australian Forestry* 77, 51-57.

Starck, W. 2014 "Are academics above the law?" *Quadrant Online*, 21 March 2014. http://quadrant.org.au/opinion/doomed-planet/2014/03/academics-law/

State Forests of NSW 1995 *Urbenville Management Area: proposed forestry operations. Environmental Impact Statement.* State Forests of New South Wales, Pennant Hills.

State Water nd *Koondrook – Perricoota Flood Enhancement Work.* http://www.statewater.com.au/current+projects/Environmental+projects/Koondrook-Perricoota+Flood+Enhancement+Work

Steffen, W., Crutzen, P.J., McNeill, J.R. 2007 "The Anthropocene: Are humans now overwhelming the great forces of nature". *AMBIO: A Journal of the Human Environment* 36, 614-21. http://www.bioone.org/doi/abs/10.1579/0044-7447(2007)36%5B614:TAAHNO%5D2.0.CO;2

Steinbauer, M.J., Sinai, K.M.J., Anderson, A., Taylor, G.S., Horton, B.M. 2015 "Trophic cascades in bell miner-associated dieback forests: Quantifying relationships between leaf quality, psyllids and *Psyllaephagus* parasitoids". *Austral Ecology* 40, 77-89.

Stokes, K., Ward, K., Colloff, M. 2010 "Alterations in flood frequency increase exotic and native species richness of understorey vegetation in a temperate floodplain eucalypt forest". *Plant Ecology* 211, 219-33.

Stone, C. 1999 "Assessment and monitoring of decline and dieback of forest eucalypts in relation to ecologically sustainable forest management: a review with a case study". *Australian Forestry* 62, 51-8.

Stone, C., Kathuria, A., Carney, C., Hunter, J. 2008 "Forest canopy health and stand structure associated with bell miners (*Manorina melanophrys*) on the central coast of New South Wales". *Australian Forestry* 71, 294-302.

Stone, C., Spolc, D., Urquhart, C.A. 1995 *Survey of crown dieback in moist hardwood forests in the central and northern regions of New South Wales State Forests (psyllid/bell miner research programme).* Research Paper No. 28. State Forests of New South Wales, Sydney.

Strzelecki, P.E. 1845 *Physical Description of New South Wales and Van Dieman's Land, accompanied by a Geological Map, Sections and Diagrams, and Figures of the organic remains.* Google Books. https://play.google.com/books/reader?id=ftUKA AAAIAAJ&printsec=frontcover&output=reader&hl=en&pg=GBS.PR1

Sturt, C. 1833. *Two Expeditions into the Interior of Southern Australia During the Years 1828, 1829, 1830 and 1831: With Observations on the Soil, Climate and General Resources of the Colony of New South Wales, Vol. 2*. Smith, Elder and Co./ University of Sydney Library, London/Sydney, Etext 2002-2003.

Sturt, C. 1838 Letter to Governor Gipps. In *Government Gazette* New South Wales, November 21 1838.

Sydney Harbour Federation Trust 2004 *Phytophthora and Vegetation Dieback in Sydney Harbour's Bushland*. Web page.

Tasker, E.M., Dickman, C.R. 2004 "Small mammal community composition in relation to cattle grazing and associated burning in eucalypt forests of the Northern Tablelands of New South Wales". In *Conservation of Australia's forest fauna*. Second edition. Ed. D. Lunney. Royal Zoological Society of New South Wales, Mosman, Australia. pp. 721-40.

Taylor, K., Barber, P., Hardy, G., Burgess, T. 2009 "*Botryosphaeriaceae* from tuart (*Eucalyptus gomphocephala*) woodland, including descriptions of four new species". *Mycological Research* 113, 337-53.

Tench, W. 1793. *A complete account of the settlement at Port Jackson including an accurate description of the situation of the colony; of the natives; and of its natural productions*. G. Nichol and J. Sewel, London, England. http://adc.library.usyd.edu.au/data-2/p00044.pdf

The Colong Foundation for Wilderness Ltd. 2012 *Submission No. 317*. General Purpose Standing Committee No. 5. Parliament of New South Wales, Sydney.

The Examiner, 1944 "Red rains in Tasmania". In *The Examiner*, Launceston, 18 December 1944. http://trove.nla.gov.au/ndp/del/article/91394829

Thoreau, H.D. 1860 *The Succession of Forest Trees*. The University of Adelaide, 2014. https://ebooks.adelaide.edu.au/t/thoreau/henry_david/succession-of-forest-trees/

Tindale, N.B., Birdsell, J.B. 1941 *Tasmanoid tribes in North Queensland : results of the Harvard-Adelaide Universities Anthropological Expedition, 1938-1939*. Unpublished Book. Held at South Australian Museum, Adelaide.

Tng, D.Y.P., Williamson, G.J., Jordan, G.J., Bowman, D.M.J.S. 2012 "Giant eucalypts – globally unique fire-adapted rain-forest trees?" *New Phytologist* 196, 1001-14.

Toohey, P. 2004 "The last of the nomads". In *The Bulletin*, 4 May 2004. pp. 28-35.

Townsend, T.S. 1846 Letter to Surveyor-General. In *Researches in the southern gold fields of New South Wales*. Ed. W.B. Clarke. Reading and Wellbank, Sydney. p. 228.

Tozer, M.G., Keith, D.A. 2012 "Population dynamics of *Xanthorrhoea resinosa* Pers. over two decades: implications for fire management". *Proceedings of the Linnean Society of New South Wales* 134, B249-66.

Tucker, A.L., Bailes, A.S., Baker, T., Burton, J.B., Ham, D., Kerr, D., Outtrim, A.R., Sergeant, C., Turner, G.J. 1899 *Third Progress Report of the Royal Commission on State Forests and Timber Reserves. The Red Gum Forests of Barmah and Gunbower: Their Resources Management and Control.* Robt. S. Brain Government Printer, Melbourne.

Turner, J. 1984 "Radiocarbon dating of wood and charcoal in an Australian forest ecosystem". *Australian Forestry* 47, 79-83.

Turner, J., Lambert, M., Jurskis, V., Bi, H. 2008 "Long term accumulation of nitrogen in soils of dry mixed eucalypt forest in the absence of fire". *Forest Ecology and Management* 256, 1133-42.

Turner, P.A.M., Balmer, J., Kirkpatrick, J.B. 2009. "Stand-replacing wildfires?: The incidence of multi-cohort and single-cohort *Eucalyptus regnans* and *E. obliqua* forests in southern Tasmania". *Forest Ecology and Management*, 258, 366-375.

U.S. Geological Survey 2005 *Pre-1980 Eruptive History of Mount St. Helens, Washington.* Fact Sheet 2005-3045. http://pubs.usgs.gov/fs/2005/3045/

University of Western Sydney 2013 *Bugs that ate a fragile woodland.* http://www.uws.edu.au/__data/assets/pdf_file/0010/582166/Riegler_grey_box_dieback_FINAL_with_image.pdf

University of Western Sydney nd *Research: Inspiring People.* Cardboard folder and leaflets. Hawkesbury Institute for the Environment, University of Western Sydney, Richmond.

USA Today, Friday, July 12, 2013. p. News2A.

Vertessy, R.A., Hatton, T.J., O'Shaughnessy, P.J., Jayasuriya, M.D.A. 1993 "Predicting water yield from a mountain ash forest catchment using a terrain analysis-based catchment model". *Journal of Hydrology* 150, 665-700.

Vertessy, R., Watson, F., O'Sullivan, S., Davis, S., Campbell, R., Benyon, R., Haydon, S. 1998 *Predicting water yield from mountain ash forest catchments.* Industry Report 98/4. Cooperative Research Centre for Catchment Hydrology.

Victoria 1939 *Report of the Royal Commission to inquire into The causes of and measures taken to prevent the bushfires of January, 1939, and to protect life and property and The measures to be taken to prevent bush fires in Victoria and to protect life and property in the event of future bush fires.* T. Rider, Acting Government Printer, Melbourne.

Viereck, L.A. 1966 "Plant succession and soil development on gravel outwash of the Muldrow Glacier, Alaska". *Ecological Monographs* 36, 181-99.

Walker, J., Thompson, C.H., Fergus, I.F., Tunstall, B.R. 1981 "Plant succession and soil development in coastal sand dunes of subtropical eastern Australia". In *Forest Succession: Concepts and Applications*. Eds. D.C. West, H.H. Shugart, D.B. Botkin. Springer-Verlag, New York. pp. 107-31.

Wallace, A.R. 1858 *On the Tendency of Varieties to depart indefinitely from the Original Type*. February 1858, Ternate. http://www.indiana.edu/~koertge/H205c/

Wallis, A.R. 1878. *Redgum. Report of the Secretary for Agriculture on the Red Gum Forests of Barmah and Gunbower*. John Ferres, Government Printer, Melbourne.

Walsh, F. 2013 *Waru, kuka, mirrka wankarringu – lampaju (Burning, bushfoods and biodiversity)*. https://www.youtube.com/watch?v=nKLD9LrySeU

Ward, D.J., Lamont, B.B., Burrows, C.L. 2001 "Grasstrees reveal contrasting fire regimes in eucalypt forest before and after European settlement of southwestern Australia". *Forest Ecology and Management* 150, 323-29.

Wardlaw, T.J. 1989 "Management of Tasmanian forests affected by regrowth dieback". *New Zealand Journal of Forest Science* 19, 265-76.

Waters, M.R., Forman, S.L., Jennings, T.A., Nordt, L.C., Driese, S.G., Feinberg, J.M., Keene, J.L., Halligan, J., Lindquist, A., Pierson, J., Hallmark, C.T., Collins, M.B., Wiederhold, J.E. 2011 "The Buttermilk Creek Complex and the origins of Clovis at the Debra L. Friedkin Site, Texas". *Science* 331, 1599-1603.

Watson, D.M. 2002 "Effects of mistletoe on diversity: a case-study from southern New South Wales". *Emu* 102, 275-81.

Watson, D.M. 2009 "Parasitic plants as facilitators: more Dryad than Dracula?" *Journal of Ecology* 97, 1151-9.

Weismann, A. 1893 *The Germ-Plasm: A theory of heredity*. Charles Scribner's Sons, New York. On-line Electronic Edition, Electronic Scholarly Publishing. http://www.esp.org/books/weismann/germ-plasm/facsimile/

Wentworth, W.C. 1813 *Journal of an expedition across the Blue Mountains, 11 May-6 June 1813* by *William Charles Wentworth*. Manuscript and Transcript, State Library, New South Wales. http://www.sl.nsw.gov.au/discover_collections/history_nation/exploration/blue_mountains/wentworth/index.html

Weste, G. 1974 "*Phytophthora cinnamomi* – the cause of severe disease in certain native communities in Victoria". *Australian Journal of Botany* 22, 1-8.

Weste, G., Marks, G.C. 1987 "The biology of *Phytophthora cinnamomi* in Australasian forests". *Annual Review of Phytopathology* 25, 207-29.

Westoby, M. 1984 *Constructive ecology: how to build and repair ecosystems*. 2nd Sabath Memorial Lecture, April 1984. Division of Australian Environmental Studies, Griffith University, Brisbane. AES Working Paper 1/84.

White, J. 1790 *Journal of a Voyage to New South Wales with sixty-five plates of non descript animals, birds, lizards, serpents, curious cones of trees and other natural productions.* Project Gutenberg of Australia eBook. http://gutenberg.net.au/ebooks03/0301531h.html

White, T.C.R. 1986 "Weather, Eucalyptus dieback in New England, and a general hypothesis of the cause of dieback". *Pacific Science* 40, 58-78.

White, T.C.R. 1993 *The Inadequate Environment: Nitrogen and the Abundance of Animals.* Springer-Verlag, Berlin

White, T.C.R. 2006 "The world is green". *Journal of Vegetation Science* 17, 539-40.

Williams, B. 2011 "Water bird numbers take off as wetlands fill Lake Eyre and Murray-Darling river basin after floods". *The Courier-Mail*, Brisbane,19 December 2011. http://www.couriermail.com.au/news/national/bird-numbers-take-off-as-wetlands-fill/story-e6freooo-1226225288284

Williams, D. nd *Book Reviews. Losing ground: Grazing in the Snowy Mountains 1944-1969. By John Merritt.* eContent Management. http://www.e-contentman-agement.com/book-reviews/review/490/losing-ground-grazing-in-the-snowy-mountains

Williams, R.J., Wahren, C.-H., Bradstock, R.A., Muller, W.J. 2006 "Does alpine grazing reduce blazing? A landscape test of a widely-held hypothesis". *Austral Ecology* 31, 925-36.

Windschuttle, K., Gillin, T. 2002 "The extinction of the Australian pygmies". *Quadrant Online*, 1 June 2002. https://quadrant.org.au/opinion/history-wars/2002/06/the-extinction-of-the-australian-pygmies/

Wittkuhn, R.S., McCaw, L., Wills, A.J., Robinson, R., Andersen, A.N., Van Heurck, P., Farr, J., Liddelow, W., Cranfield, R. 2011 "Variation in fire interval sequences has minimal effects on species richness and composition in fire-prone landscapes of south-west Western Australia". *Forest Ecology and Management* 261, 965-78.

Wolfe, E. 2009 *Country Pasture/Forage Resource Profiles: Australia.* FAO. http://www.fao.org/ag/agp/AGPC/doc/Counprof/Australia/australia.htm

Worboys, G. 2003 "A report on the 2003 Australian Alps bushfires". In *Newsletter* No. 15, June 2003. Australian Institute of Alpine Studies. pp. 1-5.

Yamaguchi, Y.T., Yokoyama, Y., Miyahara, H., Sho, K., Nakatsuka, T. 2010 "Synchronised Northern Hemisphere climate change and solar magnetic cycles during the Maunder Minimum". *PNAS* 107, 20697-702.

www.ingramcontent.com/pod-product-compliance
Lightning Source LLC
Chambersburg PA
CBHW070545270326
41926CB00013B/2212